EMANUEL SWEDENBORG

THE GENERATIVE ORGANS

CONSIDERED

ANATOMICALLY, PHYSICALLY AND PHILOSOPHICALLY

\Elibron Classics
www.elibron.com

Elibron Classics series.

ISBN 1-4021-8107-8 (paperback)
ISBN 1-4021-2815-0 (hardcover)

This Elibron Classics Replica Edition is an unabridged facsimile of the edition published in 1852 by William Newbery, London.

THE GENERATIVE ORGANS.

LONDON:
PRINTED BY WALTON AND MITCHELL,
WARDOUR ST., OXFORD ST.

THE

GENERATIVE ORGANS,

CONSIDERED

ANATOMICALLY, PHYSICALLY AND PHILOSOPHICALLY.

A POSTHUMOUS WORK

OF

EMANUEL SWEDENBORG,

LATE MEMBER OF THE HOUSE OF NOBLES IN THE ROYAL DIET OF SWEDEN, ASSESSOR OF THE ROYAL METALLIC COLLEGE OF SWEDEN, FELLOW OF THE ROYAL ACADEMY OF SCIENCES OF UPSALA, AND OF THE ROYAL ACADEMY OF SCIENCES OF STOCKHOLM, CORRESPONDING MEMBER OF THE IMPERIAL ACADEMY OF SCIENCES OF ST. PETERSBURG.

TRANSLATED FROM THE LATIN

BY

JAMES JOHN GARTH WILKINSON,

MEMBER OF THE ROYAL COLLEGE OF SURGEONS OF ENGLAND, ONE OF THE SURGEONS TO THE HAHNEMANN HOSPITAL, AND AUTHOR OF "THE HUMAN BODY AND ITS CONNEXION WITH MAN."

LONDON:

WILLIAM NEWBERY, 6, KING STREET, HOLBORN.

1852.

Published by the Swedenborg Association.

ADVERTISEMENT.

A PARTICULAR account of the MS. volume from which the following Treatises are taken, has been given by Dr. Tafel in his edition of the originals.* In executing the translation we have aimed to represent the author's meaning as closely as possible; but there are passages here and there where the sense is doubtful; and in these we have been literal, as the nearest approach to being faithful. The reader will occasionally find difficulties in his way; and he will kindly recollect, that he is reading a manuscript which there is reason to suppose was a first draft, and which would certainly not have been issued by Swedenborg in its present unfinished state. For this reason also the reader will be indulgent to the translator.†

Yet the work as it stands is a worthy integrant part of that extraordinary series of works which more than a century ago

* Eman. Swedenborgii, Sacræ Regiæ Majestatis Regnique Sueciæ Collegii Metallici Assessoris, Regnum Animale Anatomice, Physice et Philosophice perlustratum. Cujus supplementum sive Partis Sextæ Sectio Secunda de Generatione, de partibus Genitalibus utriusque Sexus, et de Formatione Fœtus in Utero agit. E Chirographo ejus in Bibliotheca Regiæ Academiæ Holmiensis asservato, nunc primum edidit Dr. J. F. I. Tafel, Philosophiæ Professor et Regiæ Bibliothecæ Universitatis Tubingensis Præfectus.

Eman. Swedenborgii, Sacræ Regiæ Majestatis Regnique Sueciæ Collegii Metallici Assessoris, Regnum Animale Anatomice, Physice et Philosophice perlustratum. Cujus supplementum sive Partis Sextæ Sectio Prima de Periosteo et de Mammis agit. E Chirographo ejus in Bibliotheca Regiæ Academiæ Holmiensis asservato, nunc primum edidit Dr. J. F. I. Tafel,

† In the first instance we had intended to publish with this volume a copious Index like that in our Translation of *The Animal Kingdom;* but upon second consideration we found that the work was too fragmentary to warrant such an addition to the size and cost of the present volume.

appeared in Latin, and which within the last ten years has been coming forth in the English tongue. What its precise merits may be, we will not prejudge; *that* is a question which belongs to the future. We see in it great intuitions of order, with a most ingenious application to details: much that is as new to the human mind now, as when the manuscript was written. We see in it also a constant amalgam of physics and metaphysics, like what there is in the human body itself; but which we do not know where to find in any author but Swedenborg. And moreover we recognize in it an affinity to Man, an addiction to central truths and principles, which is too absent from the corresponding works of this age. Yet we own that it is worth but little as a handbook for the kind of information now sought in the medical schools.

In truth the work is non-medical: it is one of those productions, which must exist more and more in all departments, and which are designed to promote a non-professional, public, or universal view of the matters in hand. Science in its universals is no tradesman, and works not for the improvement of any calling; but solely because truth is good. Such science for the human body has been cultivated by the non-medical Swedenborg.

It is a delicate subject which the present treatise embraces, but it is one which cannot be unknown. When we are little boys and girls, our first queries about our *whence* are answered by the authoritative dogma of "the silver spade:" we were dug up with that implement. By degrees the fact comes forth. The public however remains for ages in the silver-spade condition of mind with regard to the science of the fact; and the doctors foster it by telling us that the whole subject is a medical property. Swedenborg wants to tell us on the other hand all about these mysteries; and we suppose the time has come when we may begin to know. There is nothing wrong in the knowing; and though the passions might be stimulated in the first moments by such information, yet in the second instance they will be calmed by it; and ceasing to be inflamed by the additional goad of curiosity and imagination, they will cool down under the hydropathic influences of science. Well stated knowledge did never yet contribute to human inflammation; and we much question whether the whole theory of the silver spade be not a mistake; and whether children should not be told the truth from the first; that before desire and imagination are born, the young mind may receive in its cool innocency the future objects of powers and faculties which are to be subject afterwards to such strong excitements. Vegetable generation supplies a beautiful and most decorous set of analogies for instruction in animal and human. But we repeat, it will not be the great doctors as such, but the great educators, from whom this information suited to the public, and the children, can be obtained.

For the rest, the present treatise shines for us with the clear mild genius of Swedenborg. We have so often written our best about him, that we feel, were we to expatiate here, that we should be travelling upon the same lines, and weary the reader with our repetitions. It requires new voices from time to time to illustrate matters and men of world-wide importance. We are senile in the present service, and have already passed on to fields where we feel a little younger.

But with our last literary accents we would fain claim the attention of the new men of this age, to what there is in Swedenborg's scientific works, accordant with their own necessities and discoveries. In particular we suppose that there is no writer before or since who has treated as he has done, of the continuity of the body on the one hand; or of the permeation and penetration of vibrations and living influences through it, on the other. Let us take a common example. A man catches cold: straightway he feels stiffness and pains in every joint of his body: his whole head is sore; his nose runs with serous defluxion, &c., &c. Now, strange as it may appear, the present science does not present any physiological knowledge of what these pathological states may be. What is the condition of his periosteum, of the sheaths of all his stiff muscles, and of his creaking joints? How does it all happen? Neither science nor imagination knows. The feelings of the patient have no commerce with the skill of the doctor. This demonstrates at any rate that the science which lies at the basis of pathology is not yet opened. Pains, aches, swellings, and symptoms generally, glide along the body by terribly broad bridges of structure of which the anatomist wots not. Well then, there is wanted somebody besides this prim anatomist, to unfold the case. Our Swedenborg, Licentiate of No College, is one of the men in whose works we have found a beginning of instruction on this subject. He has wonderfully indicated to us many of the great bridges and highways of vibrations and influences, and in so doing has thronged with living states and forms parts which were previously dispersed, lying in sand heaps of cell germs. To the new pathology, which chronicles the passage of states through Man, he is as yet the most important contributor from the physiological side.

It gives us pleasure to end these brief lines by recording publicly that the Royal Academy of Sciences of Stockholm, the body of which Linnæus and Berzelius were *alumni*, has lately paid a fitting tribute to the memory of Swedenborg. We excerpt the following from the official account of their last annual festival.

"1852. The Academy has this year caused the annual medal to be struck to the memory of the celebrated Swedenborg. It represents Swedenborg's image on the obverse: over it his name: under it *Nat.* 1688, *Den.* 1772. On the reverse: a man in a dress reaching to the feet, with eyes unbandaged, standing before the temple of Isis, at whose base the goddess is seen. Above it: TANTOQUE EXSULTAT ALUMNO; beneath: MIRO NATURÆ INVESTIGATORI SOCIO QUOND. ÆSTIMATISS. ACAD. REG. SCIENT. SVEC. MDCCCLII." The eulogium on Swedenborg was delivered by the President of the Academy, General Akrell. It is to the liberality of this same Academy, who lent us the original manuscript, that the reader is indebted for the possession of this and many other of Swedenborg's works. Justice compels us to add that a munificent Englishman, Joseph Senior, Esq., of Dalton, impressed with the value of these works, has defrayed the entire cost of the present translation.

London, May 16, 1852.

INDEX.

PART I.

THE MALE GENERATIVE ORGANS.

PART II.

THE FEMALE GENERATIVE ORGANS.

PART I.

THE MALE GENERATIVE ORGANS.

CHAPTER I.

THE SPERMATIC ARTERIES.

1. WINSLOW. "The spermatic arteries arise most commonly from the . . . aorta, . . . about an inch below the renal arteries. Their origin however varies. Thus I have observed them to arise from the renal artery; and sometimes higher, lower, or more laterally than usual; moreover, each artery occasionally has several origins. They run down obliquely in the posterior part of the abdomen, and in the cellular tissue of the peritonæum, passing insensibly from behind forwards; and parting gradually more and more from the aorta, they cross in front of the ureters, and run through the openings or rings of the abdominal muscles, to gain the elongations or productions of the cellular portion of the peritonæum. They are very small at their origin, and in their course downward they give off pretty considerable lateral ramifications to the membrana adiposa and peritonæum, and also to the mesentery, where they seem to communicate with the mesenteric arteries. They sometimes pass through the areolæ or meshes of the spermatic veins; and before they go out of the abdomen they divide themselves into fine twigs, more or less serpentine, which are almost parallel to each other. Afterwards they insinuate themselves into the cellular productions of the peritonæum, which serve them for sheaths. They do not fluctuate indifferently from one side of these sheaths to the other; but are attached along the inner surface of the sheaths by thin membranous laminæ, which are likewise continuations of the cellular tissue of the peritonæum. They continue the same winding course within these sheaths, passing before the vasa deferentia, which are likewise contained

in the sheaths; and at length they terminate by ramifications on the testes and epididymis." (*Exp. Anat., Tr. du Bas Ventre*, n. 469—475.) The spermatic arteries are commonly two in number, sometimes more. They . . . arise . . . between the two mesenteric arteries . . . They send in the first place small branches named arteriæ adiposæ, to the common membrane of the kidneys. . . . They give several considerable branches to the peritonæum. . . . They likewise send small arteries to the ureters. . . . They communicate with a branch of the external iliac artery. (*Ibid. Tr. des Artéres*, n. 219—222.) The spermatic veins accompany the arteries, and have nearly the same course. The right vein arises commonly from the trunk of the vena cava, . . . though I have sometimes observed it to spring from the union of the right renal vein with the vena cava, and sometimes I have seen three veins on the right side coming separately from the trunk of the vena cava. The left spermatic vein arises usually from the left renal vein. In their course downward, they first join the arteries, and together with them enter the cellular productions of the peritonæum, to which they are connected in the same manner as the arteries. . . . A little below the place where they cross the ureters, they send out a considerable branch, which afterwards divides into two, one of which communicates with the capsular vein, and the other frequently with the renal. . . . They differ from the spermatic arteries, not only in being larger and their coats being thinner, but also in being more divided and multiple as they descend to the rings of the abdominal muscles; and as they produce a bundle of ramifications which becomes larger and larger as it proceeds, the ancients gave to them and to the arteries, the name of vasa pyramidalia. These ramifications communicate with each other often in this course, and form a great number of areolæ, contortions and convolutions, so as to represent a kind of plexus, which is connected to the cellular sheath of each side, by very fine laminæ. . . . These frequent convolutions gave rise to the name of vasa pampiniformia, formerly given to these vessels; and their adhesions to each other in certain places caused it to be believed that there were real anastomoses between the artery and the vein. But this does not appear to be borne out by facts." (*Ibid., Tr. du Bas Ventre*, n. 476—481.)

2. Heister. "The blood-vessels of the uterus are tortuous, and form a thousand anastomoses with each other: in women with child they dilate into a kind of sinuses. They open by little mouths into the uterus and vagina, and are the sources of the menstrual discharge. The arteries of the womb are, 1. The spermatics from the aorta. 2. Very large vessels from the hypogastrics. And, 3. Others from the external hæmorrhoidal. These all communicate in a surprising manner with one another; so that if they be inflated, or mercury or wax be injected into any one of them, they are all filled; even those on the opposite side to that on which the injection is thrown in. The veins of the uterus are also threefold, and have the same names as the arteries. They have no valves, and are much larger than the arteries, especially in women with child. Air may often be thrown into the cavities of the uterus and vagina, by inflating these vessels: and on the contrary, the latter may often be inflated by air thrown into the vagina and uterus." (*Comp. Anat.*, n. 236.)

3. "The spermatic arteries do not always arise from the aorta, but the left spermatic occasionally comes off from the emulgent, and both the right and the left sometimes proceed from the hypogastric arteries. This can scarcely have been noticed by those who say that the spermatic arteries are sometimes altogether wanting. . . . Lealis and Schraderus will have it, that the spermatic arteries, before reaching the testes, anastomose with the veins; but this is not borne out by the testimony of others." (*Ibid.*, not. 25.)

4. Boerhaave. "The seminal or spermatic arteries, arising with the greatest constancy from the front of the aorta, below the renal arteries, (though sometimes, perhaps, improperly, the left seminal or spermatic is said to come from the left renal,) are not vessels of large size. Hence, and by reason of their oblique descent from the aorta, they can hold but little blood. They run obliquely downwards, and join at an acute angle with the spermatic veins of the same side, of which the right comes from the cava under the renal vein, the left from its renal vein directly. Thereafter, both arteries and veins are included and as it were blended into one common membranous sheath or capsule, by which they are defended as they run under the inner layer of the peritonæum, over the psoas muscles and the ureters.

At length they reach their place of exit in the groin. Here the fleshy fibres of the transverse muscle lie upon them, and those of the inferior oblique, leaving a little interspace between them, transmit these vessels into their sheath. The part through which they pass in the transverse muscle, is the highest, the orifice in the oblique ascending muscle being lower down. Then about three lines lower still, they again pass through an oval aperture in the tendinous part of the oblique descending muscle; and lastly the sheath with its vascular contents, is carried over the os pubis and into the scrotum; and out of its vessels and membranes makes the testis. . . . Throughout this course the spermatic arteries give off small lateral twigs here and there. Add to this that the three several apertures of the muscles through which the sheath passes, each bestow upon it a fine membranule; but where the group of veins [vessels] approaches the testicles, the former constitute the corpus pyramidale. In its passage the spermatic artery winds a little in a spiral manner, and gives off twigs, which run in a straight and open course, are pretty ample [in proportion to the parent vessel], and derive the arterial blood laterally by true anastamoses into the accompanying vein. This is especially the case in the corpus pyramidale, where the artery near the testicle gives a branch to the lower and inner part of the epididymis, which branch spends certain twigs on the testicle itself, and afterwards terminates on the nervous tunic. Sometimes also the artery sends another branch to the top of the testicle, and constantly a large branch runs to the upper and bulkier part of the epididymis, where it ramifies and is distributed in all directions. Not to mention numerous offsets from the larger trunk, closely interwoven with the little veins in the pyramidal body, and afterwards spread over the compass of the testicle.—The pyramidal body—*corpus varicosum* or *pampiniforme*—arising from the back of the testicle, consists of innumerable veins communicating with each other, and which form a kind of net, and at length unite, and terminate in a single seminal vein. (*Inst. Med.*, n. 641—643.)

5. SCHURIG. "Poupart makes mention of a girl seven years old who had no spermatic arteries or veins. Several other writers give similar accounts. The same thing has been noticed

in the dog, hare, and wolf. Some authors tell us of four spermatic arteries, and some have even enumerated seven, &c. (*Spermatologia*, p. 43—47.)

6. "The Psoas muscle occupies the lumbar vertebræ at the posterior part of the ileum, all the way towards the anterior part in the direction of the femur. It is attached above to the last dorsal vertebra and to all the lumbar vertebræ, and it descends obliquely over the ileum, uniting at length with the iliacus. It serves to bend the femur upon the pelvis, by drawing the latter forwards. It can also move the pelvis."

7. THE *wonderful manner in which all things in the animal body are disposed and arranged, is evident on slight reflexion from the situation of the parts as corresponding to their uses. In the highest place resides the brain, which is meant to be sentient of all things, and to govern all, and which therefore is proximately surrounded by the sensory organs, and distributes the motory organs round the whole corporeal machine. Next underneath it the lungs and heart have their place in the body, upon the diaphragm : these organs put motion into all the parts in particular and in general, enabling the animal frame to live in motion. Next under these, beneath the partition of the diaphragm, in the cavity of the abdomen and within the peritonæum, are planted the viscera that nourish all things in the body, sustain continually and provide for all, and especially for the blood, so as to keep it in perpetual subsistence up to the mark of its first existence. Below this region again, and outside the peritonæum, the members devoted to generation and the propagation of the species, have their allotted places. Thus as all have their offices, so have they their destined provinces. The whole machine—the universal microcosm or little world—is provided moreover with extended arms, with loins, legs, feet, and soles, to capacitate it for application and motion according to all the circumstances that the use and necessity of the before mentioned parts can possibly require; that is to say, to enable it to obey the brain in all respects; to enable the man to live in society as a member or part of the state: to enable the body to adapt and to rouse itself for the several matters or movements that the heart and lungs require; for those that the viscera of the abdomen or the functions of nutrition demand; and for those also which the organs of procreation require, towards and in the generative act. All those actions which regard the before-named uses, are the primary actions of the appendages of the body,* i. e., *of the arms, loins, legs, and feet; all the rest are secondary, and exist inas-*

much as they regard the primary, but not as they regard the secondary acts. Thus the primary actions of the loins and feet, are meant to enable the body to take the steps, and to make the due approaches, for the act of coition, with a view to the propagation of the species; also to enable it to seek, to apply, and take to it food and aliment for the sustenance of life; likewise to minister to the diverse workings of the lungs; and likewise again in all particulars to second the commanding brain, which regards the general society. Other offices, which do not regard these directly, and which are contrary to this use, we term secondary; these may indeed be undergone, not however for their own sakes, but because, for the sake of the uses before mentioned, we are gifted with the faculty of performing them.

8. *The propagation of the human race is one, and the principal one, among the primary uses and ends; for by means thereof not only mundane society, but also the society of souls in heaven, is renewed and kept up. With this view, not only are organs of wonderful structure provided, but also the affection of love,* (*venery, as it is called from Venus,*) *is given, which surpasses all other affections; viz., to excite us to follow this end by a certain most pleasant and most delicious violence and necessity. And from this necessity it follows, that the several means that regard and promote this end, are largely and wisely provided to the full measure required.*

9. *Among the means, the blood undoubtedly takes the lead, being itself the very magazine and storehouse from which the seed is prepared. The blood is the parent of all the humors of its body, containing, as it does, the simplest elements of every degree, by the subordination and due coördination whereof, whatever of humor or essence is possible in this sublunary sphere, may come into existence.* See Parts I. and II. For besides the elements and principles of all parts of the mineral and vegetable kingdoms, it also contains the principles of the animal kingdom, that is to say, the animal spirit, in which the life and soul abide: See Part I.; also our psychological observations respecting the animal spirit and the soul in Part VI., or the last part of the work.

10. *Now in order that the blood,* (*which is in fact the parent of the seed as well as of the several principles that must lie in the*

seed from its earliest stages,) may be properly conveyed to the preparing organs, as the testes, epididymedes, uterus and Fallopian tubes; the following conditions are required. 1. *Only the purest blood, cleansed from every excrementitious quality, must have admission thither, and moreover, it must be blood that is resolvable in the most perfect manner into its elements.* 2. *The quantity of such blood must not be greater than the particular state and the natural necessity require.* 3. *And there must be an ever new stock, and an increased fresh supply during the act of coition.* 4. *This blood must not be hindered in its afflux, but be supplied with certainty.* 5. *And hence it must be conveyed to its destination by the safest way.* These are the points that we must aim at, and keep sharply in view, during our examination of the spermatic vessels, if we wish to explore the use of those vessels in a true manner, and not to attach our ideas to foreign objects that have no proper relation to the end.

11. *With regard to the first position, it is clear from very many considerations, that none but the purest blood runs to the generative organs, for the spirituous essence of the blood to be extracted, and spent upon the seed; and that hence this blood is of the softest kind, and readily resolvable into all the elements and principles. For the spermatic vessels, be they more or be they fewer, come off from the aorta immediately under the emulgents; and the emulgent or renal vessels are those which draw off all the unclean and obsolete part of the blood, and throw it in the form of urine into the kidneys and bladder, and finally out of the body: thus the blood is cleansed in the upper part of the aorta, before it is delivered to the spermatic vessels.* According to the description prefixed to the Chapter, the spermatic vessels are brought from the aorta about a finger's-breadth, more or less, below the renal, which are the emunctory vessels removing the scoriæ of the blood: in some subjects however, they arise from the renal vessels themselves, as observed by Heister, and also by the older authorities; in others again they come from the hypogastric and hæmorrhoidal vessels. From the latter origins, it may indeed be thought that their blood is not purified by the emulgent arteries, for the purpose of being supplied in a cleaner state to the spermatic vessels; but then it is to be observed that if the latter do arise from the emulgent or any other artery, still, in their course and during

their long passage, means of expurgating the blood are abundantly provided, and these means are the more considerable and numerous in proportion as the vessels under discussion spring from places in which the blood is rendered impure; which indeed accounts for the ramifications of the spermatic artery, and the derivation of the blood into these or other arteries to be mentioned presently: from this principle, no doubt, the cause of the varieties in this respect should be deduced. *Nevertheless it is a general rule, that the beginnings and middles have to convey and supply such blood as the extremes spend and require; and not only the quality of blood they consume, but the quantity also: thus the emulgents take none but the more impure blood, which has a urinous quality; but the testes and epididymides, the fixed, pure and readily soluble blood; so also the cerebrum, and the rest of the viscera.* Thus it amounts to the same thing whether the spermatic artery come off from the emulgent, or from the aorta, or from the hypogastric artery, for the testicles command the quality, and the proper quantity. For suppose that a large quantity of the purer blood is attracted in the extremes, in this case, if the purer blood is taken in, a portion corresponding must be restored by the more distant arteries leading to the part. And the same rule holds with the grosser blood also*. This disposition depends on the organ itself that consumes or expends the blood. If we carry out a number of tubes from a common channel, and draw off by one a quantity of water, but a smaller quantity from the others, still according to the attractive power exerted upon the water at the extremities, is the quantity that runs out from the common channel or

* It is by this attraction that a part of the forces of the circulation is now commonly accounted for. See Carpenter's *Physiology*, passim. Each part of the body exercises an attraction for the blood that is adequate to nourish it. This part pulls towards it this blood by the appropriate channels. Arrived at the place required, the nutritive materials and faculties are taken and exhausted; and then at once the attraction ceases. The next portion similarly attracted pushes onwards and away that which has been "used up;" and hence, by this living magnetism of demand and supply, the bodily fluids are kept in constant movement; on the one side always coming where they are required, and on the other, always going whence they are not required. The two directions are in fact one and the same; for *want* is a magnetism that lives down all the lines of the body; and that which is exhausted for one part, is *wanted* as a full object for the satisfaction of the parts succeeding it.—*Tr.*

spring. And so it is also with the blood as regards its quality. Nature, which ever strives after equation, cannot be equated justly if any particular quality of blood is taken in by larger than the proper quantities, and hence she supplies a quantity equal to the necessity. From this requirement and dispensation it follows, that the intermediate vessels are in this way necessarily disposed to pass onwards distinct sorts of blood, and no others; or if others, they send them off in new directions. This is what we term the equation of the blood, and this it is from which all the above effects result. See Part I.

12. *It is necessary that the spermatic artery should come off from the aorta by a small stem or narrow orifice, in order that the purer and more* soluble blood may be derived into the testes and epididymides; and also in order that the blood may be purified on the way, that is to say, that as it runs onwards, the more worthless, watery and urinous part, and the excrementitious portion, may be completely detached, and none but the purer part remain, which will be available for the preparation of the seed.* Observation shews that the spermatic vessels arise from the aorta by a fine and narrow beginning; likewise that the artery descends by a long course towards its proper organs; to enable it to be duly expurgated from heterogeneous matters on the passage; for the longer the distance, the greater the quantity of expurgation†. At first it throws on one side the grosser, then the purer heterogeneous elements; first the urinous, then the aqueous, and then again the oily matters that are not suitable for the seed: this is the reason why the spermatic vessels transmit branches towards the kidneys and the suprarenal capsules, and moreover to the adipose membrane both of the skin and kidneys; and indeed a number of twigs to the peritonæum, whose cellular tissue receives the aqueous portion of their blood. By this means the urinous matter is evaporated through the skin (see above), or sweated off. And finally those crudities that from other places are sent away to the liver, are committed to the mesenteric arteries; besides which the spermatic vessels communicate with the

* By more soluble blood, the author means the blood which is *more readily resolvable back into its elements;* the excellence of all supplies depending on their capability of detaching the precise materials which are wanted.—*Tr.*

† This is a rule following from the now adopted principle above mentioned.—*Tr.*

iliacs, and throughout their course with their accompanying veins. Thus whatever is binding and cramping in the blood is rejected; all those elements, we mean, in the blood, which are not soft, pliant, and soluble, but hard and unfit for its uses. Hence all the channels for discharge and depuration are sufficiently provided. Add to this that such and such blood is necessarily sent away through these channels, according to the rule just explained, whereby the extremes demand their precise quantity of blood. So it is with the emulgent arteries, the peritonæum, and adipose membrane; so also with the mesentery and the liver, all of which command their own supply. For wherever blood of a similar sort is flowing, by whatever windings it passes, it is still conveyed to its destination, because the whole system, in the interest of the equation we have mentioned, concurs to these effects or ends. *The viler part of the blood being thus cast away, the purer remains, but in order that none but the purest may enter, the spermatic vessels divide at last into capillaries and threads, and after in this wise splitting, they pass to the corpus pyramidale or the tubular offset of the peritonæum. For the finer the branches into which an artery divides, or the purer the threads that the blood traverses, so much the more purified must the latter be, and so much the clearer from its heterogeneities.* This is an evident truth reposing upon eye-witness. None but the pure, true, and genuine blood can run through the finest vessels. If anything angular, gross, or hard were to get in with it, of course the hair-tube would be blocked up, and the current or fluxion would be stopped. This was excellently observed by Leeuwenhoek, who observed only pure blood in the finest capillaries, without the multiplicity arising from volume; thus he saw in some cases the solitary globules, in others the same divided into their principles, and splitting up into six globules (see Parts I. and II.) The case would be otherwise if the artery had a thick coat. Hence the sundering of the artery into such vast numbers of ramifications, is a clear proof that the little branches receive and transmit the purer blood only. And as this division of the vessel takes place here at its extremity, it follows that the blood coming up hither, that is to say, to the neighbourhood of the testes and epididymides, is in a sufficiently purified state for entering those organs. *The neces-*

sary conditions of the higher purity of the blood are, that it be soft, thin, and resolvable into all its elements; hence abounding in spirit, and enabled to bring out this spirit from its bosom with the utmost readiness. Such is the blood which is required by the seed, which is pregnant with much spirit, and abounds moreover with the purest elements, which lend that spirit every assistance, enabling whatever in it lies to begin to arise from first principles.

13. *And moreover the utmost prudence of forethought is shewn in the spermatic vessels, to ensure a sufficient quantity of blood being carried to the organs, but not more than reason and necessity would have; to ensure, that is to say, there being neither too much nor too little; and moreover to provide that all the blood that passes in may be turned to seed. For the spermatic vessels grow not merely from one but sometimes from several origins; arising by one, two, or more trunks from the aorta, and also from the hypogastric and hæmorrhoidal vessels.* Anatomy shews that the spermatic vessels are made not only by the aorta or the renal artery, but occasionally also by the hypogastric and hæmorrhoidal arteries. Mention is also made of a branch from the epigastric vessels accompanying them, and at length at the end of their course, twining in with them: see Winslow. It is also said that they receive numerous supplies from the hypogastrics. Some anatomists again have discovered two, three, four or five branches from the aorta coalescing into a single spermatic trunk. In other subjects no spermatics at all have been found, even in persons who had had children, in which cases no doubt the spermatics came not from the aorta but from the hypogastrics. We now see at any rate that the due quantity of blood can never fail, nor the generative faculty be endangered; for when one source is blocked up, the supply comes through some other way which is open. *A quantity of the blood that has once been purified must always be ready, nor slip away before it has done its work; and for this end care is taken to prevent the pampiniform vessels, or the arteries, from communicating with the veins within the sheath or production of the peritonæum, or in the corpus pyramidale, and the purest arterial blood from slipping away by the channel of the veins.* The blood which runs here is purified, and apt for generating the seed; it is soft, and relatively pure

and soluble. But if in this place it were allowed communication with the venous blood, the intention of nature might be frustrated; Leal Lealis seems indeed to have attempted to contravene her decrees by his experience, but Winslow, Heister, and others have denied his grounds. *For the communication of the arteries with the veins takes place above these ramifications, but not within them. Below however, upon the testis and epididymis, it seems probable that there is a communication not only between the arteries and veins, but also between the veins and arteries;* as in the intestines: see above. This has not indeed been hitherto observed, yet we may conjecture as much from the connexion of things, as well as from the fact that precaution is required to prevent the quantity of blood of good quality failing; and also from other signs, and from the circumstance that the little veins are more numerous and larger than the arteries, which would not be the case if all the venous blood without loss of time were derived away to other parts; and from this further, that these veins are absolutely without valves, which enables the blood to run back into these extreme parts, and to wait in readiness, as the state and condition demand. Thus the experiment of Leal Lealis seems readily enough to meet the fact, that the blood might pass into these venous vessels by injection through the arteries, or from above into the veins, or below into the [same.] The blood therefore of the better sort, is dispensed without any loss. But on these subjects more will be seen in the Chapters on the Testicles and the Uterus. *Hence it follows that an ever new store, and an additional supply ever fresh, runs abounding in the very act of coition.* This must happen if the veins communicate with the arteries close to the testes and epididymides; but otherwise a quantity of the viler blood might run to the spot, and not only defile the seed but also burst the small vessels and capillaries. Furthermore, above this splitting of the vessels into threads, there seems to be a kind of place of reserve, small though it be, for the arteries swell out a little below their origin.

14. *One little artery communicates with another here, as in the body universally; whereby there is nothing proper in any one vessel, even the very least, that is not common to all. And this*

indeed contributes mainly to the equation of quantity and quality. So too the little veins mutually intercommunicate.

15. *This blood is unimpeded, and thus is conveyed with certainty and by the safest way; as follows from the facts, that it is not brought thither from a single source but from several; and that the vessel applies itself to the peritonæum, and adheres to it by little leaflike productions, and afterwards taking on a pampiniform or pyramidal shape, is surrounded with a particular sheath, or production of the peritonæum, whereinto the vessels do not pass stragglingly, but become adherent to the walls of this sheath.* In this way of course they are perfectly protected and safe in their course, fixed and completely attached, as they are, to the peritonæum. *Moreover lower down they connect themselves with the os pubis as a fulcrum, before they run to their respective organs.* This indeed is a provision of the utmost importance in the act of coition, when all the muscles both of the loins and thighs, and of the region under the diaphragm, and the vertebral muscles also, are in the fullest stretch of work and effort. At such a time unless these vessels were defended on the way by a common tunic, they might easily be exposed to mischief; but when a sufficiency of this kind of padding is interposed to maintain them unhurt against all dangers, then none but a general action results to the peritonæum, and from the peritonæum to the sheath or production which encloses these capillary vessels; and then the action is not perilous, although made up of the extraordinary actions of such numerous muscles. *And the common motion of the peritonæum conspires, which coincides with the respiratory motion, consequently with the motion of the nervous system from the beginning onwards, or from one end to the other; wherefore, as the motion of the blood, that is to say, the common motion of the arteries, accords with the very motion of the nerves, so immunity or safety attends the vessels. That the motion of the peritonæum coincides with that of the lungs, may be seen above in the Chapter on the Peritonæum. We all experience the fact in our persons in the alternate movements of the respiration.* All the muscles of the abdomen swell up in inspiration; the lungs acting on the abdominal viscera by means of the diaphragm, and likewise on the peritonæum itself by means of the muscles.

Moreover the form of fluxion of the pampiniform vessels concurs to the same end; for they make oval and spiral twists, like as do the minutest vessels everywhere that carry the purer blood. This form of fluxion meets the case of the most perfect fluids, which tend with safety and certitude to their destinations when they are determined in such a form. Respecting the form of this fluxion, see our former Parts throughout, and Part V. on the Doctrine of Forms. This form of fluxion is a sign of the purity of the fluid that follows it; for the purer fluids flow by their proper, supremely natural, and spontaneous path, when they are flowing in this kind of gyre.

16. *Moreover the spermatic or seminal vessels in their descent, not only apply themselves closely to the peritonæum, but also to the psoas muscle, and to the transverse muscle of the abdomen; and at length pass through the aponeurotic ring in the oblique descending muscle. Now the action of these muscles, if compared with the action of the lower body during coition, coincides so marvellously with the position of these vessels, that at this time the latter pour on more blood. For under the circumstances of venery, the action of these muscles alternately draws open and constricts, stretches and relaxes, those vessels, obliging them to emulge a large supply of blood from their springs.* This may be specifically demonstrated, only we require a knowledge of the muscles and their particular actions. The psoas arises from the last dorsal vertebra, and all the lumbar vertebræ, at the back of the ileum, and as far forwards behind the front of the femur; then it strengthens itself with the iliacus; and its action is to bend the femur upon the pelvis, drawing the latter forwards. Something similar to this occurs in the abdominal muscles before mentioned. It would however be tedious to describe the act of coition, and to deduce the fitness between the muscles and vessels from such a description; the subject too is obscene, and I pass it by.

17. *As to the venous blood, it undergoes nearly the like fortunes with the arterial. The impurer and grosser portion of it is discharged into the emulgents; but the better and thinner portion is sent towards the finer ramifications, namely, into the peritonæum and the mesentery; and this, according to the successive origin of the veins from the spermatic artery. The grosser parts first are sent away; afterwards in regular order the purer.* The

causes are similar that govern both sets of vessels. No determination of any artery takes place, or of any vein, without an end in view. Nature has an end in the least steps of all things; and hence it is that the reason of every thing may be found out: but this subject does not belong to the present Part.

18. *From the foregoing considerations we may infer in some slight degree, what are the causes of excessive salacity and pruriency, and also of excessive chastity, and of natural impotence; we may also learn somewhat of the quality of the seed as depending upon the undue and the excessive, or the over-sparing use of venery; and we may gain instruction respecting the state of the womb, and of the spermatic vessels during pregnancy: as well as respecting several other subjects which are here suggested. But these topics can be only imperfectly discussed on the ground of the spermatic vessels alone: they receive a clearer and better explanation from the several genital organs all together. For the cause comes not simply from the vessels, but also from the existing state of the before-mentioned organs. However, all changes of state in the organs themselves, appeal at last to these spermatic vessels.* As for example in pregnancy, when these vessels swell out to form a kind of sinuses. They are moreover different in the two sexes respectively, running in women to the uterus, in males, to the testes and epididymides; in our consideration of which organs we shall resume the thread here begun. *Meantime, in salacious people the spermatic vessels are larger than common, and spring from more numerous sources; the pampiniform vessels also are more numerous, complete, distinct, and swollen, and do not communicate so continuously with the veins above and below the corpus pyramidale; besides which the blood is more abundant, and more fully charged with spirit; for example in young persons: hence the accumulation of seed, and the wantonness which is felt.* Cases of salacity are plentiful and familiar enough. The mode in which the habitual use of venery excites an often-recurring wantonness, is plain from the general rule previously laid down, viz., that the use of the extremes* commands both the quality and the quantity of blood. Thus the habit of suckling causes the frequent flow of large quantities of milk; and the same

* By the *extremes* are meant the terminal parts of any series.—(*Tr.*)

rule is exhibited in innumerable other cases. See above. *On the other hand these vessels during over much chastity, or too sparing a use of venery, are obliterated, and sink away to nothing, particularly the pampiniform vessels, which are kept open by use and temperate indulgence, the flow of blood through them maintained, and the state renovated. In default of this temperate indulgence the blood is discharged into the veins, or into some other arteries, (for several places of discharge are provided); thus the spermatic vessels are unused to take their course, they become crammed with impure blood, or their walls grow together, and their calibre diminishing, they begin to perish.* Hence we have impotence, which may be either natural or acquired, and may also arise from excess and over-indulgence, in which case these vessels collapse, after having been too much emptied out; or they are unduly turgid, and at length become clogged, so that the impurer blood blocks up the pores of the testicles; for when the blood has lost its spirit, and still is allured time after time by these vessels, the latter cannot fail to lose their natural state.

18. *On the same grounds also we may judge of the quality of the seed; for although the seed is elaborated in the testicles, yet the material which forms it is taken out of the spermatic vessels. In fact, in the matter of origin, both the quality of the seed, and the capacity of performance in the generative act, depend upon the balance of the supply, plentiful or defective, of blood from the spermatic vessels, and above all, from the pampiniform vessels in the corpus pyramidale. For there it is that the blood is purified, which is afterwards prepared into seed: therefore the first seat and station of Venus is in the corpus pyramidale:* although indeed from this point there is an ample course before it, and a wide field; but the corpus pyramidale is the starting-post; a long career must be traversed between it and the goal, where the race is done and the game terminates.

CHAPTER II.

THE TESTICLES, EPIDIDYMIDES, ETC.

19. Winslow. "The ancients named the testicles didymi, or twins. They are about the size of a pigeon's egg, of an oval shape, but a little flattened on each side. We have to consider in each testicle two extremities, two edges, and two sides. . . . At the upper edge each has an appendix, called epididymis, together with which the testicle is involved in several coverings; and both testicles are suspended in a common covering called the scrotum. Each testicle is a spermatic gland, formed of a great number of very fine whitish tubes, folded and twisted in different ways, and distributed in different fasciculi between membranous septa; the whole being surrounded by a very strong common covering, named tunica albuginea. These septa are disposed longitudinally, approaching each other on one side, and divaricating on the other. They approach each other along one edge of the testicle, and terminate in a long narrow whitish body, as in a kind of axis. From thence they divaricate in a symmetrical manner, and are fixed by their opposite edges in the inner surface of the tunica albuginea, of which they appear to be a continuation. This white body may be termed the nucleus of the testicle. From this description we see that these septa are not all of equal breadth; that the intervals between them are in some measure triangular; and that the extent of the small tubes which lie therein must be very considerable. They have been reckoned to amount to many ells, by taking the sum of their several portions; and they may be easily unfolded by a long maceration, which destroys the delicate substance by which all their folds and convolutions are connected and tied down. All these small canals seem to terminate by a smaller number

of common trunks at the white body or nucleus already mentioned; which trunks afterwards pierce the upper part of the anterior extremity of the testicle, and are disposed in several folds along the lateral external part of the upper edge, all the way to the posterior extremity. From this union arises a long whitish plaited fasciculus called epididymis. . . . The epididymis is a production or process of the testicle, or a kind of accessory testis. . . . It is not of equal volume throughout; but is more contracted in the middle than at the extremities, by which it is closely united to those of the testicle. In the interval between its extremities it does not immediately touch the testicle, but is only loosely connected to it, by the duplicature of a very fine and almost transparent membrane, which serves as a kind of ligament. This membrane is the continuation and duplicature of the tunica albuginea . . . which having supplied the place of a ligament to the epididymis, afterwards invests it. The epididymis is flat, slightly concave on the under side, or that next the testicle, irregularly convex on the upper side, or that turned from the testicle; and these two sides are distinguished by two angular edges; by the innermost of which it is connected to the testicle, in the manner already mentioned, but the outer edge and flat side are loose and free. The anterior extremity or head of the epididymis arises from the testicle; the posterior extremity or tail, which likewise adheres very closely to the testicle, is incurvated from behind, forward, and a little upward, and contracting by degrees forms a particular canal, which is the vas deferens. (*Exp. Anat., Tr. du Bas Ventre*, n. 482—492.) The tunics or coverings of the testes are . . . three in number; the muscular tunic named cremaster, and the tunica vaginalis and albuginea. The first two are common to the testicle and to the cord of spermatic vessels, but the third is proper to the testicle alone. The tunica vaginalis is the most considerable of the three . . . It is a continuation of the sheath of the spermatic cord . . . The sheath having reached as low as the testicle, appears to divide into two laminæ, the inner of which forms the bottom of the sheath, and the outer expands round the testicle, and gives it a coat, called vaginalis from the Latin word vagina. The ancients termed it elytroides, from a Greek word that signifies the same thing. The inner surface of this coat is lined by a fine membrane, which strengthens the bottom of the sheath, and forms a kind of diaphragm, which prevents all communication between the sheath of the spermatic cord and the tunica vaginalis of the testicle. The cremaster . . . is a thin muscle or fleshy plane, which runs down round the sheath of the spermatic cord, and terminates in the tunica vaginalis testis. It surrounds almost the whole bag, and afterwards expands on the upper and outer part of the tunica vaginalis, where it is inserted and lost. It arises partly from the ligamentum Fallopii, and partly from the lower edge of the internal oblique muscle of the abdomen; and on this account, it seems sometimes to arise from the spine of the os ilium. . . . It is covered by a very fine cellular membrane, detached from the outside of the aponeurosis of the obliquus externus round the opening called the ring. This membrane is lost in the cellular substance of the inside of the dartos. . . . This muscle is not always red, . . . (*Ibid.*, n. 513—522.) There are two particular nerves that accompany the spermatic cord; one comes from the lumbar nerves near the anterior spine of the os ilium; it makes a curve as it runs out of the abdomen through the muscles, and in its passage serves to distinguish the cremaster. The other of these two nerves comes from the renal plexus. (*Ibid.*, n. 587.)

20. Heister. "The testicles or didymi. . . . their coats are three: the cremaster or elevator muscle of the testicle: the tunica vaginalis, which loosely surrounds the testicle, and is from a process of the peritonæum: the tunica albuginea, which is strong, closely connected with the substance of the testicle, and receives the spermatic vessels, and transmits them to the testicle. . . . The nerves of the testes are from the plexuses of the nerves of the pelvis, and from the lumbar plexuses. Numerous lymphatics may be seen in the part in a living animal. The substance of the testicles is vascular, consisting of the most minute vessels, which are the seminal vessels, and which are convoluted after the manner of intestines. . . . The testicles have apparently no glands in them. . . . The corpus Highmori in the human subject is at the back of the testis, but in the dog in the middle part; . . . it is said to have a cavity for receiving the seed. . . . The two epididymides or parastatæ . . . are oblong, almost cylindrical parts, lying on the upper border of the testicle,

and each having somewhat the appearance of a caterpillar or silkworm. They are connected to the testis by the membrana albuginea, and also to the vas deferens. . . . They begin in the testicle by five or six small seminal vessels; they end at the other extremity of the testicle, where the vas deferens begins. They are covered by a robust membrane, continuous with the tunica albuginea. They are of a vascular substance, like the testicles themselves; but the vessels which compose them are not only more conspicuous than in the testicles, but their tubular nature can be evidently shown by injections. All these vessels end at length in a single duct, which is the vas deferens. The epididymides have vessels from the spermatic vessels, the same as the testicles; nerves too, from the same branches as the latter. (*Comp. Anat.*, n. 224, 225). Morgagni shews that the first of the old authors who indicated the true vascular substance of the testicles, were Arantius, Cabrolius, and still more clearly, Riolan; De Graaf however demonstrated the connexion and use of these vessels. This vascular substance is well seen in the tortoise, the pig, the goat, and the dormouse." (*Ibid. nota* 25).

21. BOERHAAVE. "The cremaster muscle . . . draws up, sustains, compresses and expresses* the tunica vaginalis and the testicle. . . . The artery near the testicle gives a branch to the lower and inner part of the epididymis, which branch spends certain twigs on the testicle itself, and afterwards terminates on the nervous tunic. Sometimes also the artery sends another branch to the top of the testicle, and constantly a large branch runs to the upper and bulkier part of the epididymis, where it ramifies and is distributed in all directions. Not to mention numerous offsets from the larger trunk, closely interwoven with the little veins in the pyramidal body, and afterwards spread over the compass of the testicle. . . . These arteries, penetrating the nervous coat, gradually divide into branches, and taking up distinct places, split at length into numberless twigs, and these into the finest capillaries, which seem indeed infinite in quantity, being distributed over nearly the whole of the testicle. . . . Perhaps . . . they have no corresponding veins, but terminate in excretory vessels. (*Inst. Med.* n. 641, 642, 644). The testicle

* *Expresses* in the sense of squeezes out the juices from.—*Tr.*

has also small veins and lymphatics innumerable, which bring back from the body of the testicle the humor that remains after the seed has been secreted: and it seems probable that when the seed is prepared in the testicles, and not evacuated, a considerable, and a subtle portion of it, is re-absorbed by these vessels, and mixed with the venous blood in the corpus pyramidale, thereby altering the whole economy of the body." *(Ibid.,* n. 647).

22. MORGAGNI. "On one occasion whilst I was gently drawing away the epididymides of the land tortoise from the testicles, (which appeared to me, differently from what other authors describe, to be made up not of round glands, but of the finest convoluted vessels,) I saw up against the light very minute and pellucid vessels passing straight from the whole side of the testicle to the side of the epididymis; and from that time I suspected, that seminal tubules issued from the middle of the testicle as well." *(Advers. Anat.,* IV., *Anim.* II).

23. SCHURIG. "Many instances are recorded of castrated persons experiencing lust, and shedding large quantities of seed. *(Spermatologia,* p. 386—400). Ruysch says that the substance of the testicle consists of nothing but various fascicles of vessels of a blue color. . . . Instances are given of cartilaginous and osseous testicles. In some subjects only a single testicle has been found: such persons are called monorchides. In some, one testicle lies in the groin, the other in the scrotum, and then the latter is sometimes drawn back into the groin or into the abdomen. In some subjects again the two testicles grow together into one. There are other cases in which both testicles are absent; this is sometimes congenital; sometimes the result of gangrene, violent accidents, surgical operations, castration. In some people three, four, or five testicles have been present; and the same in animals. The left testicle is commonly the bigger of the two. Singular varieties in size are also occasionally noticed: cases being recorded of testicles of extraordinary bulk, and on the other hand of uncommon smallness. Calculi are also now and then found in the testicles." . . . *(Ibid.* p. 53—77.)

24. VIEUSSENS, in his *Neurographia Universalis,* plate xxix., shews the fourth nerve of the sacrum giving off branches to the

muscles of the anus, and to the glutæus major; the main branch running at length to the testes [or as Vieussens himself says, sending fibrils both to the common and proper coats of the testes]. In the way this fourth sacral nerve forms two unions; first with a branch of the third sacral nerve; and then a lengthened contact with a nerve running down from the middle of the crural nerve, and which is distributed particularly to the penis and its muscles. Lower down a branch of the crural nerve goes to the scrotum.

25. *It is commonly supposed that the testes are the only organs that prepare the semen; this however arises from not knowing what the semen is, or the essences of which it consists. The semen is not a humor of simple composition, but a humor compounded in regular order and succession; first of the inmost and purest substances; next of others less simple and pure; lastly of outermost substances, which with due prudence involve and hold the above recondite contents. The outermost in fact so involve the innermost, that in the uterus they again evolve and put them forth in almost the same order as that of their involution. The inmost parts that lie in the seed are by no means visible to our eyes, but shew and declare themselves by their effects alone. Even the outermost are scarcely apparent in their details unless under a microscope, and however high the magnifying power may be, it will still hardly shew more than the surface of parts infinite in number.* To this end it is that so many organs are constructed for preparing this single humor, to wit, the testicles, epididymides, vesiculæ seminales, prostate glands, the urethra also with its glands; every one of which contributes its share. Of all these parts we have to treat in regular succession; and afterwards we have to reconcile them with the nature and faculty of the seed.

26. *The inmost essence of the seed, which is the inmost of the man, or of whatever animal the seed belongs to, and which therefore is the life's partner, is not prepared in the testicles, but is drawn out in abundance from the blood, and from the juice of the nervous fibres. Thus the testicles are the organs that extract and supply the inmost essence of the seed, which is the only essence that makes over the principles and beginnings of things to the future offspring. Hence without the testicles there may indeed be seed, possibly similar to genuine seed in external appearance, yet*

no prolific seed. This is shewn to be the case in instances where the testicles have been lost, either by castration, accident, disease, gangrene, or owing to any similar cause. In persons thus deprived, although all the other generative organs may remain, and they may suffice to pour out large quantities of seed, still the latter is not prolific or fruitful. The humor that springs from the testicles is never presented visible externally, but only that which at length collects in the vesiculæ seminales, prostate gland, and glands of the urethra, whose humor has all the appearances of seed. *The innermost element in any humor, or in any part of any humor, is that which must first be supplied, in order that the other elements in succession may accomplish their purpose. And this innermost element must be enveloped in kindly and most corresponding essences; because the inmost is likewise the purest; and unless at once, and at the very outset, it be encased in other essences less volatile than itself, it is easily dissipated, and the principal cause thus vanishing away, the primary effect itself falls to the ground.* This first and inmost essence in every animal kingdom*, is so highly volatile, that unless it were at once imprisoned by other elements of a more fixed nature, there would be no help for its flying away; on which subject see our Psychology. And when it does fly away or vanish, still the humor left behind looks all the same to the eye; as do the surfaces or membranes of the brain, though the cortical and fibrous substances be eaten away; or the shell of a nut when the nut is gone; and all other outsides without insides. *Nature, therefore, designing to begin from the inmost, begins from the testicles, which supply the inmost.* This is best seen from the very mode of these organs; for their structure is enough to shew us, that they supply an abundance of the purest seed.

27. *If the structure of the testicle be properly examined, it will be evident, that it is so wonderfully constructed on geometrical principles, that anything more perfect cannot be. Thus the capillaceous vessels or most minute tubules which form its texture, run in tortuous convolutions after the manner of little intestines, and aspire to a form and to determinations superior to the circle in perfection. These several vessels, thus coördinated, are connected*

* Each animal or animal body is an animal kingdom.—*Tr.*

together by the finest bands, so that every one of them, in any individual or singular effort, has regard to the common or general weal; and, in due succession, to the more general weal; and at length to the most general. And these exquisitely fine or individual bonds of the vessels, unite together to form more general partitions or septa; these septa again coalescing into one most general coat, which is the tunica albuginea. Thus there is not a part of the testis ever so minute, but stands related in regular order to the general membrane. This membrane again, by all points, has reference to a certain common axis, which is the node at one side of the testicle, or as it is called, the Corpus Highmorianum. This axis, being a congeries of centres disposed in the manner of an axis, respects its focus or extremity, as the common centre of all the determinations; that is to say, it respects in this wise the prostates or epididymides. Thus, by the mere power of form, all the several tubes of the universal testis respect the epididymis as the terminus ad quem*—the goal of their courses, whither they direct their essences, previously poured into more capacious tubes or canals, in the most exact manner possible. Thus the testicle, by virtue of its very form, performs with the utmost safety and exactitude, and so to speak, with spontaneous determination, all the functions for which it is designed. In the human testis this axis is placed on one side almost out of the sphere, but not so in the testis of the dog. This is in order that the two testicles, though disjoined, may still in a manner be one. For when each of them determines itself towards the surface, to the axis, and both axes are represented as united or single, then this is instead of a centre; and each regards one origin, one corpus pyramidale, or one spermatic vessel. These points being plain to the anatomist, and geometrically demonstrable, I shall say no more about them.*

28. *This wonderful reference of the parts to their general is exhibited in each of the testes. All this reference however, or all this determination, is peculiar to the testis itself, inasmuch as the whole of the structure that is enclosed in the tunica albuginea, belongs properly to the testis. The vessels, indeed, and fibres, are continued from the vessels and fibres of the body; but with this reservation, that on entering the body of their testis they are the body's no longer, but thenceforward belong to that new authority*

or province whereto they have been assigned. In fact, they are disjoined from the other provinces of the body, or rather from its regions, i. e. *from the body itself, by a particular septum, not unlike the diaphragm, and which is placed in front transversely just under the corpus pyramidale, and is connected on the one side of the membrane to the inmost tunic of the corpus pyramidale or production of the peritonæum; and on the other to the inmost membrane of the tunica vaginalis testis: very much as the great diaphragm partitions off the epigastric and hypogastric regions of the body. By this means the vessels and nerves that enter the tunica albuginea, are exempted from the body's sway, and are brought under that of the testicle. Yet furthermore, in order that the testicle may not be a part quite discriminated and separated from the common stock, but may be continuous from it in a general manner, the peritonæum descends all the way to the testis, and clothes it over with a coat borrowed from and continuous with itself: hence the tunica vaginalis, which is a continuation of the peritonæum, and consequently of all the abdominal viscera which are enclosed in the peritonæum. And at the same time, in order that the testis may depend upon the body itself in a most general manner, and be subject to its motion, it is still further covered by a motive or muscular tunic continued from the tegument of the body, namely, by the cremaster muscle, which is extended from the oblique muscle, as well as from its aponeurotic substance, from the corpus pyramidale or spermatic cord of vessels, and over the anterior part of the testicle: and thus it is kept in the same general stream of motion as the abdominal muscles, viz., in the pulmonic motion. In this respect, however, the testicle is only the common appendage of the lower region of the body: but to constitute it also the common appendage of the whole body, it is girded about with the dartos, bursa or scrotum, of which we shall treat further on.* The simple statement of these particulars makes it evident how nature unites together the parts and viscera of the animal frame, and also their provinces and regions; and at the same time distinguishes and separates them; for whatever part is not continued to the general or from the general, is no proper part of this kingdom: nevertheless, it is so continued, that notwithstanding, it is a member that possesses its own property and belongings. To look into all these matters distinctly would require pages and pages. But as they are superficial and formal, and do not specially concern the preparation of the seed, I do not dwell on them longer, more particularly since a similar rule of arrangement is universal in the body.

29. *With regard to the alchemical preparation of the seed in the testicles, it is among matters the most difficult of apprehension, as presupposing the knowledge of several subjects which for the sake of order and series it would not be proper to explain at present. It presupposes that we know what the soul is; what the animal spirit is; and what the blood; all of which enter into the composition of the seed. For before we can know how the parent's soul passes into the future offspring, and likewise understand the whole faculty of growing into a corresponding body, we evidently require a knowledge of the nature and faculty of the first substance, which is the formative, creative, plastic substance of the new embryonic body, and which in the purest sphere of things represents in the most perfect manner all that exists in the great sphere, and that is to exist in the new being. But as we must advance and mount analytically, that is to say, rise from effects to the region prior to effects, viz., to causes and principles, so I am bound by the case to set forth here at the threshold certain particulars which will not be rendered clear until we return to them in the sequel, and come at last to knowledge. Meantime, the end will crown the work.* Here I will only repeat, as I have already treated of the blood, that it is the complex of all the possible essences in the body; that is to say, it not only consists of vastly numerous elements and principles from the elemental world; but moreover its inmost essence is the animal spirit, and the inmost of this is the soul itself, which is the first and the proper substance of its body; from which, and for which, all the rest exist.

30. *In order that the semen may be so prepared that it can serve for the formation of the new body and new life, the inmost, purest, simplest or highest essence of the parent himself must be put into it; and this must moreover have the inmost place in the semen given to it: in this way each part of the seed cherishes within it, deeply and intimately, an element of the purest and highest kind. And in order to bring this about, the latter must*

needs be drawn out from the blood and nervous juice in the first organ, which is the testicle; and ultimately by degrees in the succeeding organs, it must be enveloped or included in other elemental substances, to prevent it from flying off in aura, and to bring it without fail into use. Thus it is that the principal essence extracted in the testicle, is afterwards enwrapped in a regular series of envelopes in the parastatæ, vesiculæ seminales, and in the prostate glands, and even the urethra. This is now to be shewn, and in the first place we have to demonstrate that the testicles pour on the true vital essence, the veriest plastic substance; for although seed is produced without the testicles, such seed is not prolific: as is abundantly shewn by cases of castration, and by injury, gangrene, ossification and old age as affecting these organs. This is the sum of our treatise; for the semen which is visible to the eye, is not immediately produced by the testicle, but something is always added to it in the other organs. The inmost and vital part is certainly absent if the testicles are wanting. Now if the seed is to be summoned into existence from its first principles, these must perforce be poured into it in the first organ or the testicle. That which is innermost must come the first in the series. We see therefore what the peculiar office of the testicles is; and this will be still more evident when we consider the structure of the testicle and the infusion of this essence. With respect to those cases in which prolification has taken place after removal of the testicles, they will be explained in the sequel, especially when we treat of the vesiculæ seminales.

31. *Those canals or intestinular tubes of which the substance of the testicles is made up, are nothing but continuations of the finest vessels from the spermatic arteries.* This is shewn by the experiments of numerous anatomists, as Ruysch, Morgagni, De Graaf and Heister. These vessels pass into the testicle through its proper coat, the tunica albuginea. *Besides these vessels continued as above, and circling about in wonderful windings, immense multitudes of nervous fibres also flow in,* which according to a custom universal in the organic frame, flow into all these vessels, both great and small, and smallest; hence in the testicle, into each*

* We have used the words *flow in* for *influunt*, occasionally, because, although not usual English, they seem to carry a meaning of influence and fluidity which belongs to the ideas of the author.—*Tr.*

and every canal; which, if we may judge from the multitude of nervous fibres, they completely envelope around, and bestow upon it a kind of new tunic. We shewed by many considerations in Parts I. and II., that the nervous fibres in the body flow almost without exception into the bloodvessels, and scarcely into any other parts. For as the organic mechanism of the body is principally called into existence by the bloodvessels, so in order that the nerves may call it into action, it is necessary that the fibres of the latter should enter into the closest relations with the vessels: hence their nervous coat, and the others, which are in fact constructed of none other than nervous principles. For the one prevailing thing in the organic body, is even the fibre, which in descending by degrees, and forming its organism, essays in the first place to inform the sanguineous canals; as may be seen proved in the Part on the Fibre. Especially is this the case in the testicle, which is entered by multitudes of fibres; for some go to it from the mesenteric plexus, and by consequence from the par vagum and great intercostal nerve, which accompany the spermatic vessels: also from the lumbar nerves; then again from the great ischiadic and crural nerve; and moreover a large branch is supplied thither from the fourth pair of sacral nerves; as will be shewn below. *As such a quantity of nervous fibre flows into the testicles, and comes closely into relation with every tubule thereof, it follows that this fibre pours in its spirit or noblest essence immediately; and indeed, the very essence that the compound fibres, thus that the most simple fibres, contain.* It is plain to be seen in all the viscera of the body, that the fibre applies itself to the vessels: so it is in the spleen, lungs, brain; also in the several costal vessels, which the fibre binds in a kind of ivied twine: see above, and the Part on the Brain. And as in these organs (the testes) the fibre winds round every tubule, and completely opens out or unfolds, (for it flows in in sufficient quantity to enable it to unfold,) so it follows that each and every simple fibre, thus unfolded and terminated, pours in here its essence, which is the innermost and first of all. See Part I., and the Part on the Fibre, where we shew that the nervous fibres pour their spirit into the bloodvessels exclusively, to the end that the blood may be impregnated therefrom, and be constituted the humor that plays the soul's part in the body. *It is then the fibre that puts the first essence of the seed immediately into the tubules, and in fact supplies that which is the vital essence; that is to say, the life itself, and the principles and commencements of the bodily life. This alone it is that can be called the plastic and prolific essence, and the principal of all: the rest however exist on its account, to enable it to build up its form aright out of their combined materials, and to commence weaving the web of the animal body.* The nature of this substance, force, or essence, will be shewn in the Parts on the Animal Spirit and the Soul. Meantime it is indubitable, that the soul which is carried over into the offspring, lies in the seed, and is not poured in at a period subsequent to impregnation. Unless this were the case, surely the female might conceive, though the testicles of the male were removed, or the semen depraved, if any seminal moisture, or anything tending that way, were sprinkled upon her organs.

32. *A mere examination of the nerves which flow in, will be sufficient to shew how great is the abundance of this essence which is poured into the testicles by the simplest* nerves. For fibres are supplied thither not only by the intercostal and par vagum, but also by the lumbar, and the great crural or ischiadic nerves; and moreover a particular branch is sent from the sacrum; thus the supply comes at any rate from all those nerves that carry the largest amount of essence.* The intercostal, and the par vagum or eighth pair of nerves, are the largest and most general nerves of the body; after which comes the ischiadic nerve, which is the biggest of all the trunks; not to mention the peculiar nerve from the sacrum, which Vieussens figures and demonstrates in his *Neurographia*. *If we well examine these several nerves it will be very evident, that all the fibre† that flows into the testicles, descends from the medulla of the cerebellum, but not from that of the cerebrum: hence that the innermost and vital essence of the seed owes its origin, not to the cerebrum, but to the cerebellum.* See Part I., where we prove, that the intercostal nerve and par vagum are nerves of the cerebellum, and not of the cerebrum. The ancient anatomists, with Hippocrates, were aware of this

* *i. e.*, by the nerves in their *ultimate or least fibrillæ, into which the author intimates that they divide themselves in the testicles.*—*Tr.*

† We here use the word *fibre* as a collective noun.—*Tr.*

fact. The same remark respecting origin may be made of the nerve from the sacrum, delineated by Vieussens (*Neurog.* tab. xxix., n. 14, 15), and which runs to the testis; for the whole residual fibre of the cerebellum, at the end of the spinal canal, comes forth at last in the sacrum: see the Parts on the Cerebrum and Spinal Marrow. A nerve moreover proceeds from the great crural, and comes into close relations with this nerve from the sacrum, and after a long contact therewith, at last leaves it. It is a doubtful point whether this nerve is the offspring of the cerebrum or cerebellum, for the great ischiadic nerve, which springs from the lumbar nerves, is spent on the muscles of the loins, legs, and feet; and therefore is of cerebral origin; but it does not run immediately to the testicles, but on the other hand comes in contact with a nerve that is determined thither, but afterwards again quits it; and thus only communicates to it the abundant spirit which it carries. Besides which a part of the fibres of the cerebellum also goes to form the above-mentioned great crural nerve. Moreover the plain fact proves, that the fibre of the testicle comes from the cerebellum, for the testicles are nowise under the dominion of the will, and cannot be called into activity at the will or disposal of the cerebrum; their excitation arises entirely from a kind of natural spontaneity and instinct. Furthermore, if the conception of this seed and of its first virtue depended on the will, *i. e.*, on the cerebrum, all this faculty would go to wreck after the briefest exercise, and the very purpose of nature in the earliest age would be quite frustrated. It follows then from the fibres themselves, from the facts of the case, and from the reasons, that the cerebellum is the butler and steward of the inner seminal essence.

33. *Whilst the sexual love is proceeding from the first effort to the final act, and when at length the state is hot, so sweet a tremble and so bland a soothing seizes upon the whole nervous system, that all the fibres whatever in the brain, and all whatever in the body, expand in every way with innermost delights, and open every way for the escaping spirit. So much so that they are all striving to lavish forth their whole spirituous treasure wherever they can get it out; and particularly do they spend it into the arterial blood, and into the testicles, in which the biggest*

nerves terminate by their ultimate fibres. Hence the lassitude that follows venery, and the debility which is felt after the excessive exhaustion of the nerves, joints and muscles, particularly those of the loins, thighs, legs, feet and footsoles. We may gather from plain phenomena, that the delicious affection of which we are treating, produces the effect of expanding the several nervous fibres; for all the pores everywhere stand open: nay, the mind too is intimately exhilarated, and feels a desire to communicate to the other party all its most precious secrets; joy naturally expands the nervous fibres; how much more so, this inmost joy, which far surpasses all joy. That highly pure tremble which is present, shews the same thing. Necessarily therefore the expansion extends to the fibre that terminates in the testicles; what happens in the act or fact is abundant proof of this. For at this time the whole of the better animal essence rushes from those grand nerves, the intercostal and par vagum, and from the crural nerve, into the testicles; whence the debility of the parts dependent on the crural nerve, and the universal lassitude in the whole nervous system. The latter however does not arise from the quantity of spirit that flows out in the seed, but from that which runs forth everywhere else in the body into the blood and into the glands. Hence the fibres, deprived of their juice all over, must needs experience a kind of lovely swoon; but from which they soon recover, for the spirit of the blood reverts back again to the fibres through the universal circle which we term the circle of life, and respecting which see Parts I. and II., and also the following Parts or IV. and V. *Thus sexual love, and particularly the act of coition, are to be considered as an effort to transfer the entire life or vital essence into another self, or into the offspring,—as a wish to transfuse all one's own life into the new life of a distinct being; and as this love springs from the inmost, the act itself expands the most simple fibres, which contain and carry the first essence of the animal spirits themselves; and this essence they breathe out not only into the testicles, but also into the blood; for the blood has the task of conceiving and preparing the other elements which belong to the composition of the seed.* The mere effects of sexual love are sufficient to prove, that it proceeds from the inmost grounds, and expands the simple fibres. For this love is of an unspeakable nature, exceeding all external delight, and therefore proceeding from the inmost. As for the fact that it involves the effort of transfusing into another being the whole of the vital essence, this too is evident from the love itself—the storgē or parental affection—which naturally begins in the act of copulation; for we love our offspring as ourselves, and more than ourselves—so much so that whatever in us lies, we long to pour forth into our progeny; but the wisest care is taken, lest the effort should be successful, and all our life be shed away, as happens in certain living creatures, and notably, in the silk-worm. *The time of youth and early manhood is the fittest for the performances of Venus, because at this period the simple fibres are naturally swollen with spirit; but this is not the case as age comes on.* The fibres of the youthful are excellently distinct; apt for agitation in fine detail; not as yet glued together; and moreover the animal spirit comes in generously from its proper sources; the blood too is full of it: hence the fibre is most prone, and the blood is most prone, at the very first idea and contemplation of the object; and the loss of the animal spirit is easily repaired: not so however in older people. *The cerebellum particularly is all effort and work, for it is the fountain head of the life of the seed; and therefore when the cerebellum is put into this state, it spends all the forces of its spirits on the fibres that cater for the genital members; and robs the other organs of their essences: hence the universal origin of that lassitude which is occasionally felt after the act.* This follows from the connexion of things, and from the facts of the case. For all parts of the frame covet rest after repeated performances, when much ardour has been felt: and this is still more the case in those who pour forth a large quantity of seed, and who are more advanced in life. Reserves however are provided in the vesiculæ seminales, (of which we shall treat presently,) to prevent this lassitude from seizing upon the nervous system and the brain. These vesiculæ are to be regarded as emunctories of the accumulated seed: thus it is not always produced immediately from the fibres in the large quantities that are shed forth. It is further to be observed, that the great crural nerve gives off from the middle of its body a particular branch, and brings this into close contact with the nerve that is sent from the fourth pair of sacral nerves to the testes: see Vieussens, *Neurog.*, tab. xxix. This gives ground to conjecture, that this nerve, by its connection with the nerve of the testicle, pours on a copious essence derived from the body or belly of the crural nerve; and that thus the testicle is never deficient in supply, being maintained from so generous a source: particularly as this nerve, the biggest in the body, is in full work and motion during the act, and lavishes all its powers; so much so that it might properly be called the venereal nerve.

34. *From the proper tunic of the testes, viz., the tunica albuginea, which is principally a nervous coat, and ultimately subdivides into septa, all those nervous fibres pass into each particular tubule; the tubules however are still but continuations* of the blood-vessels: thus the fibres are led about by a nervous membrane wherever they go; and the blood runs in a system of nervous tubes.* The tunica albuginea is also termed the nervous coat, by Boeerhaave as well as other writers: that this tunic affords a passage through the septa into the several tubules, see above. *But although the principal essence of the seed flows from the nervous fibres into these little canals, yet this does not hinder the soft and most genuine blood, but which has now undergone division into its constituent parts, and even into its animal spirit—in fine, the white blood, from running up into the same tubules; for this soft blood contains a quantity of similar essence, but enveloped in the purest elements and principles of the elemental world.* For the fact that the blood contains such an essence, which is its principal one, see Parts I. and II., and also our next Part, on the Brain, and indeed our treatises throughout; hence the blood also comes in aid. *But this blood contains that purest seminal essence already enveloped in parts belonging to the elemental world, and not therefore pure or perfectly free from earthly dregs; and hence it is sparingly supplied to the testes, and indeed in only*

* It is not here to be understood that Swedenborg affirms the literal continuity of the seminal arteries with the tubules of the testis: it is rather a passing over of the one into the other with a change of nature at the point where the one ends and the other begins. In this sense, all the hollow structures appear to end by blind extremities or loops, yet the pores in these ends are continuations of the cavities and vessels. No writer before Swedenborg had so just an idea of the discrimination between parts, as combined with their evident continuation into each other.—*Tr.*

sufficient quantity to enable it to serve for the reduction of that purest essence to its least forms, and for surrounding it with a kind of exquisitely fine membrane; the pure essence forming a sort of nucleus in the innermost, which has to be discriminated from its fellow nuclei, by the blood with its elements contributing a part in the formation. This cannot be made so clear as it ought to be, until I shall have attempted a description of the parts of the seed. But let it be premised, that the vitality of the seed requires that it shall intimately include the above pure essence in every part. In the blood this essence is not pure, and is not kept in a nucleus, but it is tied and bound up with the purest principles of the elemental world, and the animal spirit is the result. So long as this essence is thus enveloped, and not pure, that is, not left to itself, it cannot possibly set about the formation of a new body. The testicles then are provided with a view to this pure essence flowing in immediately from the fibres; only such a quantity of the blood or its animal spirits being furnished, as suffices to intimately enclose this essence, like a nucleus in a very fine shell. It is for this reason that seed cannot possibly be prepared by mere blood. *For this reason also there are not so great a quantity of arteries supplied to the testicle as to the parastatæ or epididymides; for the animal spirit alone proceeds from the compound fibres and their canals, and also from the disintegrated blood; but the true first essence of the spirits comes only from the simple fibres, of which the compound are made up.* But on these points see the Part on the Fibres, the Animal Spirit and the Soul, for unless the latter be well understood, what we have set down will be a foreign tongue. *The mere length and tortuosity of the canals mainly contributes to all being carried on in a proper manner, and according to the intentions of nature. For through these very long and very fine canals none but the purer part of the blood can be driven, the part, namely, which corresponds to the nature and peculiar elasticity of these canals, consisting as they do, of fibres. Besides which no greater quantity can be admitted of the blood that has undergone resolution into animal spirit, than the natural necessity, as measured by the use and end, demands.* But how all these matters are carried on, we can at present only guess, in the absence of positive information derived from the microscope: let us however remark, that all is done in a more distinct manner than in the large or compound canals, according to a rule laid down in Part I. *As the blood contains within it the primitive elements of every degree, and as every existence which is possible may come out of these, and furthermore as the seed, the purest essence of animal life, accumulated in the most minute tubes, requires assistance and support, to suit it conveniently and distinctly to the several parts it has to construct and play, so of necessity the blood here comes liberally in supply, and gives the aid which is wanted.* That the blood contains within it all those elements of which the entities of the elemental world are made up, see Part I. Now it is necessary that the blood be invited hither; for nothing can be formed without its ministration, since without it the instrumental cause would at once be absent, without which the principal cause cannot expatiate beyond its own sphere. *Whence it follows, that no seed can be prepared apart from the assistance of the blood and its animal spirit; hence that the first conjunction or formation is effected in the testicles and their tubules, particularly in the neighbourhood of the larger canals, or at the place where the lesser canals come together beside the corpus Highmorianum or the lateral node.* If the most pure animal essence in which the life resides were not at once fixed by means of the animal spirits extracted from the blood, its nature is such that it would instantly volatilize, and be dissipated. Hence it follows of natural necessity that this essence must be fixed before it passes into canals of any size. Nor do I think it can be doubted that an additional quantity of blood is derived thither. Such being the conditions, of course the purest substances have the inmost place; and the less pure the place next about and around the former; not only fixing, but so enfolding them as the uses to be performed require.

35. *These exquisitely fine forms of the seed, commenced in the testicles, are at length completed in the epididymides; that is to say, they are surrounded and bathed with the multitudinous spirit of the blood, and are fixed in such a manner as to embolden them to venture further outwards, and to run forth through the vasa deferentia: and therefore a larger quantity of blood is derived from the spermatic or pampiniform vessels towards the epididymides, than towards the testicles themselves.* A number of

pipes or conduits originating from the innumerable small vessels of the testes, run to the epididymides, to which an abundance of the nobler blood is supplied. The vessels themselves have been clearly seen as they pass and penetrate, and therefore no doubt can be entertained of their actual existence. *This spirit of the blood is elicited not only from the blood itself, but also from the testicles; for the fibres that flow into the testes are compound, and carry the animal spirit: this spirit cannot get into the tubuli themselves, the latter being constructed of none but the simpler fibres: and therefore, before they enter, they let this spirit drip out into the interstices between the tubules; and when there, it has no resource but to be derived towards the corpus Highmorianum or axis, and thence onwards; and also immediately into the epididymides, where it performs a similar service to that of the blood resolved into similar spirit.* Morgagni intimates that there is something conveyed from the testicles to the epididymides immediately, and not by the highway of the tubules; and this something can be no other than that mystic essence that courses within the tubules, or is sent forth by the fibres first in a compound form: that this essence is the animal spirit will be shewn in the Part on the Fibre. *Therefore in the epididymides there is no new composition of the seed whose formation is commenced in the testicles, but there is simply a condensation of that exquisitely fine envelope wherewith the true substance of the seed and the vital essence are enswathed. This little tunic, which is shapen by the animal spirit especially, at length grows strong enough to allow it with safety to be put out and extruded through the vasa deferentia into the vesiculæ seminales.* Every the minutest part of the seed may in a manner be compared to a little egg, whose materials would easily run away unless they were defended by a circular shell. This shell or rather tunicle is prepared on the passage by the corpus Highmorianum, and particularly in the epididymides. In the latter then no new composition is accomplished, but the safe keeping of this most noble essence is provided for. The first envelope that surrounds and includes it, agreeably to the order of nature, must be the substance that comes nearest in point of perfection to the essence enclosed; in a word, it must be the animal spirit; moreover also the purer elements of the elemental world, which are contained in the blood, and which thicken this fine shell, and give it immunity against all perils. Now this is indeed the very primordial operation that is performed in these organs which we first encounter; other organs then succeed them in regular order, and make up the humor that is visible to the naked eye, and which is superficially termed the seed. And it is to be remembered among the preliminaries, that the seed is resolved back again into its parts, in the womb, Fallopian tubes and ovaries, in the same order in which its composition is effected in the male organs. Furthermore, that a considerable supply of animal spirit comes in from the epididymides, may be inferred from the fact, that they too are surrounded with the tunica albuginea, consequently with a tissue of a fibrous kind.

36. *With regard to the tunica vaginalis testis, which is a production of the peritonæum covering the spermatic vessels, it appears not only to serve the purpose of protecting the testis, and of enabling the arterial vessels and the nerves to enter the part with safety under its shield, but also of receiving all that more impure matter* that the arterial vessels excrete, and which would not be conducive to the first formation of the seed.* For here, at the threshold, the arterial blood cannot be so pure but that it still has with it a certain serosity and comparatively gross material: moreover in the very globules of the blood there lie many things that are by no means good enough for this virgin and most pure seed; for instance, urinous, alkaline, saline, and many other ingredients: hence receptacles are required whereinto the heterogeneous and gross parts may be sent, and the nearest receptacle that there is, is the tunica vaginalis, which loosely surrounds the tunica albuginea, and moreover is finely cellular, descending, as it does, from the cellular tissue of the peritonæum. Underneath this coat pass the vessels of the testes and epididymides, and they give off thither many twigs and threads. Therefore this tunic is provided to absorb the grosser

* These doctrines of Swedenborg respecting the permeability of the membranes are well worth attention, particularly since similar statements come continually before us as results of the now admitted laws of endosmose and exosmose. It would appear that there are definite tides proceeding everywhere, of which probably Swedenborg was the first person to advertize us. The tracing of the fasciæ of the body will furnish important illustration of these corporeal courses.—*Tr.*

material of the blood, and to allow none but the purer essences to make their way into the laboratory. *But the outermost and muscular tunic, or the cremaster muscle, is continued from the ligaments of Fallopius and from the corpus pyramidale, or the cord of spermatic vessels: and does not seem either to compress or relax the testicle, but only to erect it, and by the erection to enable the blood to flow in from the spermatic vessels, and the spirit and purer essence by the nerves, into the testis and its appendages. For unless the testis be erected, these vessels as well as the nerves are folded together, and also the middle septum or small diaphragm, and hence an obstacle is raised against the first matter of the seed flowing in. This is the reason why the testes and epididymides are raised up and made rigid during the venereal act; while on the other hand in the intervals the connexion is more flaccid, or in a manner ceases: by this means the irritation is not so great whilst the act is not performed; and it becomes still less during seasons of prolonged chastity.* Anatomists are agreed that the cremaster raises the testis and epididymis, but does not compress them. If it did the latter, it would exert only an empty force upon the proper coat of the testis, the tunica vaginalis, which is the intermediate coat. The testis obtains its sufficient power of action from the abundance of its own proper spirit. But when erected, it is then so placed or disposed, as to admit the transflux of the blood in the freest manner, as well as the transflux of the spirit of the fibres. Its erection proceeds from the same cause as that of the penis and other generative organs, namely, from the nerves, when the brain is powerfully moved with love. For then all the spirit, exhilarated to the utmost with venereal love, opens the very minutest fibres, and thus erects the muscular fibres, and particularly the cremaster; for these very nerves also flow into its motive fibres. By means of this animal dispensation the wisest care is taken to prevent the testicles from continually alluring away a quantity of the purest blood and spirit, which would be to the injury of the whole system and animal. *It is not for a moment to be denied that the testicles enjoy a motion of their own, a kind of alternate expansion and compression, but almost imperceptible. For not only a vast quantity of fibre, but an uninterruptedness of artery, enters into their structure and pervades it. Nor likewise must we deny to them, any more than to the rest of the viscera throughout the body, an extraordinary motion, during the whole of those periods when the business of venery is going on. Nor again a general motion, evidently synchronous with that of the peritonæum, i. e., with that of the lungs..* Of the proper motion no one doubts, for though the minutest vascular twigs do not always enter, yet the larger branches do. So too with the nervous juice. But as to the position that the general motion, or that of the whole body of the testis, coincides with the motion of the peritonæum, or the pulmonic movements—this is shewn by the mere connection of the parts. Thus an offset of the peritonæum is continued around the spermatic vessels all the way to the testis, where it opens out and clothes the testicle in what is called the tunica vaginalis. So likewise there is an aponeurosis from the internal oblique muscle of the abdomen, which stretches round the external tunic. Thus when these muscles and the peritonæum are raised up, or relaxed, this very appendage itself must needs be similarly moved; and since the first motion coincides, as every one knows, with the breathings of the lungs, the general motion of the testicles must of necessity be synchronous with the pulmonic motion.

CHAPTER III.

THE SCROTUM.

37. Winslow. "The scrotum is the cutaneous covering of the testes. Outwardly, it is a bag common to both, formed by a continuation of the skin of the neighbouring parts, and commonly very uneven, having a great number of rugæ on its outer surface. Interiorly it is fleshy, and forms for each testicle a muscular capsule termed the dartos. The external or cutaneous portion of the scrotum is nearly of the same structure with the skin, . . . only it is somewhat finer, and likewise plentifully beset with sebaceous glands, and bulbs of hairs. Though it is a common covering for both testicles, it is nevertheless divided into two lateral parts by a superficial and uneven prominent line, which appears like a kind of suture, and hence is termed the raphe. This line is a continuation of that which divides in the same manner, the cutaneous covering of the penis, and is continued through the perinæum, which it divides likewise, . . . all the way to the anus. It is only superficial, and does not appear on the inside of the skin. The inner surface of this cutaneous bag is lined with a very thin cellular membrane, through which the bulbs and glands appear very distinct when we examine it from within. The rugosity of the scrotum is in the natural state commonly a mark of health. The dartos or fleshy portion of the scrotum, is a true cutaneous muscle; the fibres of which are for the most part closely connected to the skin, running through the cellular substance which lies between these two portions, and serving in the place of an adipose membrane, but without the least appearance of fat. This muscle is thin, and by the disposition of its fibres forms

a bag with two cavities, or two small bags joined laterally to each other, and contained within the common or cutaneous portion. The lateral parts of these two bags, which are turned from each other, are longer than those which are joined together. By the union of the adjacent parts of this double dartos, or of the two dartoi, a septum is formed between the testes, which anatomists call mediastinum scroti. The raphe or suture already mentioned adheres to the edge of this septum, and thereby braces down the middle of the cutaneous portion, which thence appears to have two cavities. . . . The other edge of the septum adheres to the urethra. The two bags of the dartos are lined on the inner or concave side, by a cellular substance more considerable than that between the convex side and the skin; so that the fleshy fibres, all the way to the septum, lie between two cellular beds. They run through the outer bed, as has been said, to be inserted in the skin, and by their contraction they form the natural rugæ of the scrotum. These fleshy fibres have likewise a close connection with the internal cellular membrane, especially at the upper part, below the groin, where the anterior and external lateral portions of the dartos terminate by a kind of tendinous or ligamentary expansion, which is strongly united to the internal cellular membrane. I have often shewn this, as a particular fascia lata, which gives insertion to the portions of the dartos just mentioned, and as a broad frænum which keeps the same portions together. The aponeurotic or ligamentary expansion of the dartos is fixed in the ramus of the os pubis, between the triceps and the origin of the corpus cavernosum of the same side, . . . all the way to the lower part of the symphysis pubis. The internal portion of these muscular bags, or that which forms the septum scroti, is fixed to the urethra by means of a communication between the same ligamentary expansion, and another, of which we shall speak in its proper place."—(*Exp. Anat. Tr. du Bas Vent.*, n. 493—503.)

38. Heister. "The scrotum is a little bag or pouch . . . with hairs upon it in adult subjects. We note in it the suture, which divides it into two portions, a right and a left: . . . the muscular membrane termed the dartos, which serves to corrugate it: the septum, formed by the doubling of the dartos, and

by which the scrotum is divided into two compartments. . . . Its vessels come from the pudendal and hypogastric vessels; its nerves from the sacrum." (*Comp. Anat.*, n. 224.)

39. Schurig, in his *Spermatologia*, 1720, Cap. II. § 32, p. 79, says: "The scrotum in cases of hernia, in hydrocele and sarcocele, as well as in intestinal hernia, is greatly extended beyond the normal size, and takes the shape of a huge oblong cupping glass. A case is given in which it reached the extraordinary size of a foot and a half long, and three feet and a half in circumference, and weighed thirty-four pounds*. In one case the scrotum reached the knees, and not a trace of the penis could be seen [p. 81]. In one instance it was far bigger than the human head. There are also cases recorded in which the scrotum was distended by flatus to the size of a young child's head. In Egypt the part is frequently subject to fleshy excrescences of a huge size; so much so that a piece removed from one person weighed twenty-five pounds [§ 33, p. 82]. Examples are related of the scrotum being absent, while the penes and testicles were entire, such cases are congenital, or the result of gangrene, or of exposure to intense cold, by which the scrotum has been killed by sphacelus [§ 34, p. 84]. Sphacelus of the scrotum [§ 35, p. 86]. A case of the scrotum being covered with a stony crust [§ 36, p. 87]. Cases of calculi of amazing size in the scrotum [§ 37, p. 88]. And one of a wart hanging from the part, and in shape like a mushroom.

* Dr. Esdaile's cases in India are still more extraordinary. In many of them the tumour weighed from 60 to 100 pounds. See his *Mesmerism in India, and its practical application in Surgery and Medicine:* London, 1846; and ALL his Indian labours in *The Zoist*, vols. iv.—ix.—*Tr.*

40. The *genital members occupy the fourth and lowest region of the body, and stand out beyond the third or abdominal region, being at once separated from it and united with it; a proof that all are dedicated to peculiar uses. The genitals however are in a manner the extremities and ends at which the members that stand above them, aim. The latter not only excrete in the former all superfluous and inharmonious materials; but they aim thither for the further reason, that where the end is, the beginning again of the new animal being may also be; which beginning occurs when the male is conjoined with the female, and they coalesce to make one body out of two.* For this law* of nature is of universal prevalence, that where there is an end there is a beginning, on the model of the perpetual gyre or circle of geometry; whence a kind of perpetuity and continuity arises. *In order then that all these members or organs devoted to generation, may be mutually connected, and absolutely conspire to one common end, this cavity is prepared, covered over with coats, and discriminated by a septum, by which all these organs are marvellously connected, and mutually intercommunicate. This cavity is the scrotum, divided into two compartments by the septum and mediastinum, just as the thorax or the two lungs are divided. This septum communicates with the spermatic vessels; also with the didymi or testicles and their appendages by the vasa deferentia; thus with the vesiculæ seminales; likewise with the urethra, and so with the prostate glands; which connexions, apart from the scrotum, and from the gyre that its tunics make, would have been impossible to realize.* These particulars are of a purely anatomical nature, and clear enough to all who study the science of anatomy. The connexion itself

* The law of the contact of extremes is a branch of this universal law.—*Tr.*

then is too manifest to be called into question, and I shall not further attempt to prove it. But in order to shew the reason of all the connexions, we need an explanation of the other genital members to be treated of in the sequel. Here we can only remark, that beside this septum the spermatic vessels enter the testicles, and from this septum the urethra is continued: hence, as this septum communicates with the first and last parts immediately, it also communicates with all the others mediately. *There are two orifices at which the superfluity of the whole body is excreted; there is the orifice of the urethra, through which the urine is discharged, and that of the anus, for the unloading of the fæces. These orifices serve as the extremes of the whole body. They however communicate with each other in the most evident manner. Thus externally there is a kind of white tendinous band, which, commencing from the orifice of the urethra, runs thence along the lower tunic [under side?] of the penis, and through the scrotum and the perinæum, all the way to the other orifice, viz., the anus, where it ends. Furthermore, this band is fixed to the extreme part of the septum that divides the scrotum into two compartments, and runs all the way to the insertion of the spermatic vessels, and at the same time to the beginning of the urethra. Thus it makes a sort of circle, and gathers itself up into a centre, wherefrom it again goes to the circumferences.* It is perfectly obvious, and indeed the anatomists record it, that this white tendinous band reaches from the lowest part of the gland, or from the orifice of the urethra, through the scrotum to the anus. And then again Winslow shews that this same band, whose course is only superficial, in a degree is adherent to the septum likewise. Hence we see that it tends by this septum to the spermatic vessels, and also to the beginning of the urethra; for to the latter the septum is attached: see Winslow. Thus it appears that a kind of gyre is formed, beginning as it were from the centres, and ending on both sides in the circumferences and their two foci. *And the same thing again occurs when we penetrate inwards by these two orifices, viz., the urethra and the rectum; in this case too we come to a peculiar centre, namely, to the beginning of the urethra, or to the neck of the bladder and the prostate glands; the rectum being adherent by ligaments to the neck of the bladder.* Thus a similar gyre occurs. This is the reason why these members have so close an intercourse with each other, and the penis is erected when the bladder becomes distended, particularly in the night: also why the urine is held while the fæces are discharged; and why both these evacuations are suspended when the semen is emitted. The student will best inform himself respecting this connection by dissection and plates: we shall have occasion to recur to it in the sequel. *This is the reason why the several genital members are excited to their performances immediately and mediately according to the connexion of the parts,—naturally in coition, artificially or lasciviously whenever this band or general frænum under the orifice of the urethra, or in the peritonæum itself, is softly handled or stroked.* For on both hands the effect of the titillation is propagated from the circumferences to the centres, that is to say, to the spermatic vessels, testicles, vasa deferentia, vesiculæ seminales, prostate glands, and urethra; which could not be the case were it not for the scrotum and the septum thereof, which wonderfully conjoins all the parts, and through a double gyre, external as well as internal, carries up to those centres the whole caress and tremble of the touch. By virtue of this gyre and connexion, there is not a part but at once feels a like titillation calling it into play. *Moreover, the scrotum serves not only for connecting the several generative members, but also for guarding and conserving them against all injury and loss from whatever cause, extrinsic or intrinsic. Particularly does it thus minister to the testicles, which are the chieftains of the band, and do the principal work. For the scrotum is as a cushion to the testicles, irrigates them moreover with pleasant humor, and keeps them uniformly safe in their natural state, and in their proper places.* It protects them, that is to say, against heat, cold, knocks, rubbing against the adjacent parts and tunics, &c. And to secure their indemnity, they require to be always circumfused with warmth and moisture, like the rest of the abdominal and thoracic viscera; and like the brains, &c., of which we have already spoken. *Moreover, the dartos or muscle of the scrotum, by a kind of external and general aid consequent upon the contraction of its fibres, raises the testicles into the natural situation; thereby enabling the blood of the spermatic vessels, and also the spirit of the fibres, to flow in as* *they ought, and likewise at the same moment to open the passage for the spermatic vessels and the fibres:* as we observed before, when speaking of the cremaster. Whilst it is in action, it raises the testicles, and gives the blood the opportunity of flowing in through the arteries, and the spirit, through the fibres, into the testicles. But as this force is only particular,* and very feeble, a general force is required to coöperate. For while the muscle contracts and corrugates the scrotum, and the muscle of the septum does the same, the testicles are raised up as a matter of course to the natural level by an external and general force. And in order that the cremaster or particular muscle and this general muscle may thoroughly conspire together, it sends forth on both sides, and through the septum itself, a kind of ligamentary, tendinous and aponeurotic fascia, to concur entirely with the action of this muscle. Thus the forces of the dartos or scrotal muscle are concentrated about the insertion of the cremaster. And further the dartos concurs with all the other muscles that call these genital members into exercise; but in a peculiar way, as we see from the nerve that is sent to the scrotum from the sacrum; which nerve forms a very remarkable contact and conjunction with the great ischiadic nerve, and also with the nerve that goes to the testicles; being in fact a means of union for all those nerves that excite the muscles to the act of venery, and at the same time for the cremaster muscle. The mutual relation and conjunction of the muscles, corresponds with that of the nerves, because it is the nerves alone that excite the muscles to action. See Vieussens, *Neurogr.*, tab. xxix. *This is the cause of the wrinkling or rugosity of the scrotum, to wit, that when need requires, it may put the testes, and at the same time the other genital appendages, in a proper attitude for use; and also that in the intervals when there is no such need, it may give and relax; and the seminal essence not run perpetually into the testes, depriving the blood and the other viscera of their vital spirit.* For the first or spirituous essence of the seed does not flow in, save when these members, and particularly the testes, maintain their natural position. Hence when the scrotal muscle is relaxed, and they fall out of this situation, these members are deprived

* In contradiction to general.

for the most part of the afflux of seed. Hence when the dartos contracts them, and corrugates the scrotum, it is a sign of the abundance of spirits, as well as of health and proclivity to venery, in short of sustained vigor. *For the rest, we shall have to show in the sequel, when we come to speak of the urethra, the use which is performed by the two cellular tissues of the scrotum; namely, the particular humor that they carry, and the destination whither they send it.* The cellular tissues are two in number, one on each side of the muscle; as is likewise the case in the intestines and other parts. And it will be explained further on, how these tissues receive the expressed humor from the muscles, not only from their own muscle which lies between them, but also from several others, and how they drive it into circulation, and lastly discharge it for the service of the seed.

CHAPTER IV.

THE VASA DEFERENTIA.

41. Winslow. "The vasa deferentia are two white, firm and somewhat flattened tubes, one on the right side, the other on the left. From the epididymides, of which they are continuations, . . . they run up in the cellular sheaths of the spermatic vessels, as high as the openings in the abdominal muscles; the blood-vessels on each side lying forward, and the vas deferens behind them. The bundle thus formed by the blood-vessels and vas deferens, and their common covering, is termed the spermatic cord. . . . The covering is closer on the outside than on the in, and for that reason it has been looked upon as a sheath. The internal tissue, which is more cellular than the external, connects all the vessels together, and the external forms a covering to invest them. The vas deferens having reached the membranous lamina of the peritonæum, where that lamina covers the orifice of the sheath, separates from the blood-vessels, and runs backward in the form of an arch in the cellular tissue of the peritonæum, as far as the adjacent side of the bladder. It passes afterwards behind the body of the bladder, to which it adheres very closely, as also to the lamina of the peritonæum which covers it, and then continues its arched course towards the neck of the bladder, where both vasa deferentia meet, and their arches terminate. In this course, the vas deferens crosses behind the neighboring umbilical artery; also crosses the extremity of the ureter of the same side, in its passage between that extremity and the bladder; and then, having got behind the bladder, it meets the vas deferens of the other side, between the insertions of the ureters, and the two run down together to

the neck of the bladder. This canal, which at first at its origin from the epididymis, is pretty large and plaited, soon becomes smaller and closer in texture, and so continues till it gets behind the bladder, where it again begins to be larger and irregularly plaited. It arises from the angular portion or posterior extremity of the epididymis, and from thence runs forward in a very oblique course, on the posterior half of the epididymis, where it bends a little back to mount behind the spermatic vessels. The smooth portion of this canal is firm, and as it were cartilaginous, especially round the surface of its cavity, which though very narrow, is still kept open by means of the solidity and thickness of its sides. The cavity of the vas deferens is cylindrical, though the whole tube is flat, and its external circumference oval, as may be seen by cutting it transversely. The cavity enlarges as it passes behind the bladder." (*Exp. Anat., Tr. du Bas Vent.*, n. 504—512.)

42. Heister. "The vas deferens or ejaculatory duct is a strong, whitish canal, something like a nerve, and of about the thickness of a middle-sized straw. One such vessel runs from each epididymis all the way to the vesiculæ seminales and the urethra. The course of each vas deferens is as follows. From the epididymis it ascends through a process of peritonæum to the abdomen, and is reflected thence to the neck of the bladder. It terminates in the front portion of the vesiculæ seminales, running in part straight into the urethra, and in part obliquely into the vesiculæ seminales; whereby it can deposit its liquid in either, as the case may require. In substance it is very robust and tough, strong almost as a nerve. Its cavity at the first, and indeed in its course, is so fine as scarcely to admit a bristle: but near the bladder the calibre increases considerably; and at the end of the tube again narrows, and affords no spontaneous passage into the urethra, but only opens during the venereal paroxysm." (*Comp. Anat.*, n. 226.)

43. The *vasa deferentia are the emissary vessels of the testicles considered as glands, and communicate with the latter through the epididymides, from which they immediately proceed. The axis of the testicles, or corpus Highmorianum, is continued all the way to the vas deferens; there being a continued axis reaching to the urethra, and which ultimately ends in the glans penis. The seminal vesicles do indeed take up this continued axis on the way, but they are only diverticula, and notwithstanding them, the canal is also continued into the cavity of the urethra.* The vasa deferentia come off from the epididymides, but still we may conclude from the connexion here, that they really spring by means of the epididymides from the very axis of the testicle, or the corpus Highmorianum. This axis concentrates the forces of the whole testicle on the whole epididymis; and thus by means of the epididymis and all its tissue and the whole of its fibres, causes the force to terminate in the vas deferens. On examining the epididymides we find, that the entire cavity and porosity of each stands related to this orifice alone; which therefore is a continuation of the axis, consequently of the testicle, following the directing line of all its parts. And that this axis, thus continued, is determined onwards to the urethra, appears from the immediate communication of the vas deferens with the urethra, and its mediate communication with the vesiculæ seminales, which latter are stations on the road, as is plain from their situation, and also from their office, in that they are repositories for the superfluous seed. *Hence all the fibre of the whole testicle and its canals, which is determined by the septa into the tunica albuginea, and from this into the corpus Highmorianum, is continued to the vas deferens as far as the urethra.* This is somewhat

like the case of what we term the corporeal fibres, or those which proceed from the cutaneous glands, there making a new origin: in which respect the vas deferens deserves to be called the great or grand fibre of the body.* *Hence it follows that the whole force and action of the testicle is determined at last to this vas deferens serving as an axis, hence that the vas deferens draws its proper determination from the testicle and all parts belonging to it; for no fibre enters the vas deferens immediately, but is the continued fibre of the testicle, and particularly of the tunica albuginea; hence from no other source can it receive its power of action. And as the testicles have a peculiar motion of their own, and likewise a general motion, and an extraordinary motion besides in the act of venery; so also have the vasa deferentia. The proper motion of the testicle passes into the fibres and the coat of the organ; and so too does the whole of the extraordinary motion, and the general motion also, which synchronous with that of the peritonæum and with that of the lungs, is continued by means of the peritonæum and its production, and finally by means of its bladder also.* As to the proper motion, this is plain from the connexion just pointed out, which is tightened every time the testicles are raised up by means of the cremaster muscle. Now the vasa deferentia are brought into the same situation and faculty of action at the time, for they proceed under the same capsule or sheath with the spermatic vessels, and are excited by the same common muscle. And that they are subject to the same common motion is plain enough from the fact, that they are likewise included in the production of the peritonæum, and continue their course to the aponeurotic ring of the internal oblique muscle; so that whatever general action results to the spermatic vessels, and therefrom to the testicles through their tunica vaginalis and muscular coat, the same also comes to the vasa deferentia, which at the time are kept in the like stream of motion with the most individual fibres, and which they receive from their first origin, or from the very brains. So the whole effect follows with inevitable certainty. *Now hence it follows, that there is some considerable expansion and constriction of the*

* See the Part on the Fibre, and also *the Animal Kingdom*, Part III., where Swedenborg developes his positions respecting these "corporeal fibres."—*Tr.*

vasa deferentia, brought about by the tunic itself, and which must be all the more considerable in proportion as the tunic is thicker and consists of more numerous fibres; for there is not a fibre in the vas deferens but corresponds to a fibre in the testicle. The tunic of the vas deferens is very thick and robust; of an oval form externally, but almost spherical within; shewing that it consists of the multiple and almost continued but compacted fibres of the tunica albuginea testis. Hence it is, that every fibre of the testicle stands related to a fibre in the vas deferens. Hence it follows that there is a notable expansion and constriction; arising, that is to say, from the very parts or fibres themselves, which during infancy are softer, consequently fitter and readier for action than in adult life, when they are condensed into a substance not very dissimilar to cartilage. *And in order that these vasa deferentia may be adapted to every general motion of the muscles of the abdomen, peritonæum and bladder, and may always be safe from harm whatever perils of motion surround them, they are carried round in ovals: otherwise, were they to be extended in too straight a line, the general motion would perturb the proper and interior motions of the vessel, and above all, the extraordinary motions.* We may all know by simple reflexion on our own persons, what the action of the abdomen is during the various bodily efforts: what it is for instance in struggling, in stretching ourselves in the morning, in sneezing, in discharging the bowels, and particularly what it is in the venereal paroxysm; moreover, by the same self-experiment we may observe the action of the bladder in forcing out the urine: in a word, the several actions, as well natural as extraordinary. If therefore the vasa deferentia did not run in windings and arches, but were stretched forth straight, or approaching to it, they would of course undergo extension and contraction in all these violent motions, when yet their expansion and constriction ought to depend on the action of the testicles; and hence the two motions would be confounded. Wherefore they are so adapted to the peritonæum both in the capsule and beyond it, that they can give or yield in whatever direction the force is applied, and be freely left to their own principles of motion.

44. *Now as the tortuous tubules of the testicles carry only the purest essence of animal life, and as in the grouped canals and*

finally in the epididymides an abundance of animal spirits are thrown in, which like surfaces cover in the parts of the essence as kernels, or envelope this substance in a shell of the softest kind, so now it follows that the canals, which carry this kind of substance thus composed or fashioned, are made up of the fibres of the testes, consequently are purely nervous; and that this nerve is similarly composed with the substance that flows through it, in order that the continent may exactly correspond with its content. This follows from the very composition of the seed. In its first composition it consists of nothing but the purest substance or essence of life, which the animal spirit or spirit of the blood softly surrounds in the way of a surface: of course then the tunic cannot fail to be correspondent. Every tunic is thoroughly accommodated to its contained fluid, as we observed in Part I., in speaking on the subject of *actives*. Thus the nature of the content is known from the continent; and the correspondence in the present case is verified by the examination or enucleation of the seed. *A similar work to that begun in the epididymides, is continued throughout the whole extent of the vasa deferentia; that is to say, the aim all along is, that the purest convolutions of this seminal substance may be surrounded with a fitting superficies, so that this superficies may continually increase in strength and thickness through the whole tube of the vas deferens; which is aided by the very length and perpetual whirling or circumvolution of the parts.* The first composition, as we said above, is carried on in the epididymides, yet is not completed there, but only slightly commenced. A quantity of spirit is poured on by them, which is elicited in abundance from the blood and the compound fibres. This juice or this spirit is poured copiously by the epididymides into the vasa deferentia. Thus as this genital substance runs through its canal, it is continually circumfused by the juice and spirit that environ and beset it; especially as it has to pass through so long a tract. For if the motion of every globule is to be inferred from the form of fluxion of its simple substances, it will follow from the argument of nexus or connexion, that the form of fluxion of each is of the most perfect kind, namely, is perpetual-spiral or vortical; respecting which see our Doctrine of Forms. Whence it follows, that each minute seminal or genital globule of the kind will perpetually, that is to say, by vortical whirlings, cover itself over with a new and thicker superficies; and this, all the way from the appendices of the testicles to the vesiculæ seminales or urethra. *Thus the deferent canals complete this work that the epididymides begin.*

45. *In order then that all things may be duly carried on, according to the universal order of nature, the beginning of operations and the end of the same must concur; so also the influx of the blood and spirits into the testicles, and their efflux from the testicles. For this end the spermatic vessels are enclosed in one and the same capsule with the vasa deferentia; and the two flow in and flow out at once in the same place; so while the venereal act continues, as much of them flows in as flows out; and in fact these two actions correspond or keep step.* Nature everywhere keeps and cultivates her universal laws, for they are most perfect, and breathe continuity or perpetuity. When therefore propagation and the necessity of society are the business in hand, the principles and ends are connected in no light fashion, but the connexion extends to a considerable tract; and the spermatic vessels that go into the testis, and the vasa deferentia that go out of the testis, are contained within one and the same capsule, all the way to the aponeurotic ring of the abdominal muscles; thus the effect cannot possibly cheat the cause. *And while all the fibres are put on the stretch in the venereal paroxysm, mobile and tremulous therewith, it is further necessary that the continued fibres of the vasa deferentia should be in the same state; and in fact in a still greater and grosser tremble, mounting with the degree of composition, so as to act upon the enclosed seed not only by reciprocal expansions and constrictions, but also by a certain peculiar tremble or quiver. And as the seed of its own nature is most prone to every motion, and especially to that of the most perfect gyration, so it is further carried most vehemently into the latter motion by the similar accessory power of the tunic.* These purest substances of animal life are adorned with privileges of supreme perfection. They possess the primordial and highest force or power of elasticity; the like power of expansibility, compressibility, modificability, gyrability and fluidity. Hence when external impulsive force is added to the internal, this force is so exalted, that the least impulse sets it off in efforts as it were spontaneously towards its outlets or goals, though it may

have to traverse the longest canals. Facts themselves teach us the mighty force that lies in the spirit from one extremity to the other, or from the cortical gland of the brain to the motive fibre of the sole of the foot. This genital substance consists of similar spirits, but is more compound, and thereby formed into a substance of the next lower degree; wherefore it cannot fail to have the next quality and perfection, as will be shewn in the Doctrine of Forms. *Now hence we see why so narrow a canal suffices for this substance, and through which nevertheless an abundance of genital humor may flow out.* We must not consider the capacity of the tube, without taking into account the purity, swiftness and elasticity of the fluid that runs through it; all of which are mutual correspondents. A quantity of a fluid of the purest kind, which is elastic, and rapid in motion, will run through the smallest tube, while only the same quantity of a gross, corporeal and comparatively slow humor, runs through a much larger and indeed a very considerable channel. Whoever knows anything of either geometry or physics knows the truth of this remark. *This is the reason why the vasa deferentia at the beginning and the end, are so narrow as scarcely to admit a bristle, while in the intervening course their calibre is greater; from which it may be inferred, that the orifice leading into the urethra is closed excepting in the act of venery, but the orifice into the vesiculæ seminales, open; the urethral orifice being on the other hand open during coition.* This conclusion comes not only from the fact, but also from the organic connexion of the parts; as appears when we examine closely the insertion of this tube into the urethra. Here at its beginning the urethra also is kept closed. Furthermore, the erection of the penis, and the tremulous excitation of the vas deferens, also the action of the testicle on this external part, all are efficient causes of the opening of this tube at the two ends.

CHAPTER V.

THE VESICULÆ SEMINALES.

46. Heister. "The vesiculæ seminales are two membranous, cellular receptacles, united to the back part of the neck of the bladder, and meant to receive the seed from the vasa deferentia, to complete its constitution, and at the proper time to drive it out through the urethra. They are three finger-breadths long, and one finger-breadth across. They consist of a strong, vascular and as it were muscular membrane, forming on the inside various cells which communicate with each other, and which is also enveloped in a robust layer of peritonæum. The cells however may be resolved into a single cavity, as it were a small cœcal intestine. Each vesicula has its own excretory duct, and each duct generally opens by a particular orifice into the urethra. . . . The vesiculæ seminales have very numerous bloodvessels from those supplying the bladder and rectum. Their nerves come from the pelvic plexuses. Their inner surface is reticulate and wrinkled, like that of the gall-bladder. Some authors allege having seen glands in them." *(Comp. Anat.*, n. 227.)

47. Winslow. "The vesiculæ seminales are two whitish, gibbous, soft bodies, about three or four finger-breadths long, one broad, and about the third of one in thickness. They are situated obliquely between the rectum and lower part of the bladder, in such a manner that their superior extremities are at a distance from each other, and their lower extremities united together between those of the vasa deferentia, of which they imitate both the obliquity and the incurvation. They are irregularly rounded on the upper part, and their breadth decreases gradually from thence. By the union of their lower extremities

they form a kind of fork, the branches of which are broad, and bent like ram's horns. These extremities are very narrow, and form by their union a small neck, which runs behind the bladder toward its orifice, and continues its course in the groove of the prostates, in the substance of the contiguous portion of the urethra, till its extremities pierce the substance of the caruncula. . . . The inner substance of the vesiculæ is plaited, and in a manner distinguished into several vesicular capsules by tortuous folds. Their external surface is covered by a fine membrane, which serves for a border and bridle to the folds, and is a true continuation of the cellular tissue of the peritonæum. The vesiculæ may easily be unfolded, and all their tortuosities straightened out, and by this means they become much longer than in their natural state. Their inner surface is villous and glandular, and continually furnishes a particular juice, which digests, exalts, refines, and perfects the semen. . . . The passage of the vasa deferentia into the vesiculæ is very remarkable. I have already observed that these canals are incurvated behind the bladder, and that their contracted extremities unite at that place. They unite in an angle, and run between the contiguous extremities of the vesiculæ; and this union is so close, that the adhering portions seem to form only one middle septum between two small tubes, each of which is formed partly by the extremity of one vas deferens, and partly by that of the neighboring vesicula. The lateral union of the extremities of the vas deferens and vesicula seminalis on each side, forms likewise a kind of very short septum, which terminates in a crescent, like a small semilunar valve; the extremity of the vas deferens being much narrower than that of the vesicula. By this mechanism, the fluid contained in each vas deferens has liberty to enter the contiguous vesicula little by little, but that contained in the vesicula cannot enter into the vas deferens. If we blow into one of the vasa deferentia after compressing the urethra, the air inflates the contiguous vesicula seminalis and the urinary bladder, without passing into the vesicula or canal of the other side, except we blow with violence. Afterwards the two small tubes formed each by the extremities of the vas deferens and vesicula, run in between the basis of the prostates and canal of the urethra; and perforating the sides of that canal obliquely, they terminate in the caruncula." *(Exp. Anat., Tr. du Bas Vent.,* n. 545—552.)

48. Boerhaave. "The vesiculæ seminales, connected by membranes to the back part of the neck of the bladder, are two small cœcal intestines, folded together into gyres, crypts and recesses. . . . The ejaculatory vessel . . . enters in almost a straight line into these little intestines near their most capacious fundus. Again each vesicula has another and an emissary canal which joins the former at an acute angle; and the two emissaries uniting in the urethra at an obtuse angle, terminate in a single emissary duct which opens into the urethra. And meantime both the vesiculæ are bound together by a muscular membrane."

49. Schurig. "Nature sometimes endows the vesiculæ with a great many bendings, and distributes them into a kind of venereal woof, consisting of eight or nine cells." *(Spermatologia,* p. 40, § 35.)

50. In *the course of nature, none but the superfluous part of the prime animal essence and its spirits is sent away into the seed; that part namely which is over and above after the proper quantity has been spent upon the blood, the fibres, humors, glands and members throughout the frame,—in a word, upon the universal economy. Those essences which are proper to each living creature are perpetually conceived, generated, and hatched, to meet all the cases of life, by their fountain-heads in the most perfect little laboratories of the brain, that is to say, in the cortical glands. The organs which receive this noblest superfluity are the testicles; those which store it up are the vesiculæ seminales; and those which excrete it are the several parts of the urethra; wherefore all these organs, the genitals as they are called, stand out beyond the trunk of the body, and there in the due season receive what remains over and above the outlay of the system. And these organs are formed, increase and decrease, to suit the whole course of their uses. Thus at the beginning of life, when no more is generated than is necessary for the formation, vegetation and nutrition of the growing frame, and for the economic distribution and animal functions thereof, these organs are not perfect, nor opened for the above use; but as soon as ever the higher organic body has attained a certain stated dimension, they then for the first time deposit this superfluity; as we see in middle age; in youth, manhood, and the adult period. Lastly, in old age, when the overplus ceases, and no more is produced than is fairly wanted to sustain the body, then these organs are again closed, they diminish, shrink: and indeed when a smaller quantity than is enough for the body is engendered by the proper sources, as in the various stages of decrepitude, the very size of the body is lessened. Thus all things in the system, in general, as well as specifically and particularly, are exactly*

conformed and proportioned in their habitudes to the ratio of the supply and abundance of this essence,—the genuine animal essence, the first and the veriest, which governs everywhere in its little world. The equilibrium or equation here indicated may rightly be termed preëstablished animal harmony. But of these subjects we shall speak at greater length in our Psychological Parts. *This superfluous part of the noblest animal substance, carried down, as it is, and unloaded into the testicles, is sent away by the vasa deferentia into the vesiculæ seminales; and is drawn out at length from the vesicles through the canal of the urethra; in which last emissary channel, moreover, the grossest, vilest and last dregs of the blood are evacuated. For it is a constant law of nature, that the first and the last meet, with this view, that all the intermediates or middles may fitly conspire to both the ends; for by this means the likeness of a perpetual gyre goes on in all things.* Respecting this constant law of nature, see what we have said above in various places. *But although this most noble essence of life in this way appears as if it too were excrementitious, or as if the animal world must be cleared of it as of a wealth not conducive to good, yet be it remembered that as it is the essence in which the life and first principles of the animal kingdom, and the very nature thereof in the primitive substances, lie, so it is wisely intended, that it be not sent away on the score of superfluity, but for the purpose of procreating offspring. For this reason all is so provided as that the seed does not run out without a cause, or of its own accord, but is moved forwards to its outlet by an antecedent will,—by means of incentives, and by virtue of recipient objects and subjects, offered to it, and set before it, and which complete the course of the means to the end; and thus in man it is intended to be dispensed at once rationally and naturally. Moreover care is taken, and a kind of rational instinct is employed, to prevent any other than the redundant portion of this most noble substance from being thrown forth, lest any part of the proper life necessary to every one should decrease or be lost. This is the reason why this superfluous substance does not flow into the testicles all at once, and at stated seasons only, but successively and slowly, in exact proportion and adaptation to its own generation and to the constitution of the body. And in order that this most noble extract may be drafted away successively and in similar pro-*

portion through the vasa deferentia into the vesiculæ seminales, and may there find a repository, and collect, increase in quantity and be perfected in quality, until its supply is so copious, and its character such, that it ought to be sent for this use or end; and in order that it may stay there long enough, and not burst forth or break open the doors, before the last incentive forces come: for these reasons it is superabundantly provided, that when it is emitted, it shall be primarily for the purpose of procreation, and secondarily for that of evacuation; reason and the mind being its dispensers.

51. *The superfluity of animal life, or of the essence wherein life so lies that it deserves to be itself called life, is sent away especially into the testicles, and therein is prepared until to become the inmost, highest, most perfect and principal substance of the seed, and which alone is genital and prolific. This is collected in the vesiculæ seminales, and there grows up and is finished, until by attaining the proper quantity and quality, it is fit to go forth to execute the use which nature has designed. The seed may not improperly be compared with the seed of the subjects of the vegetable kingdom. Thus the seeds of trees and plants are intimately enveloped in a kind of fruity sheath, and as soon as produced, they are at once surrounded with a tegumentary covering, to protect them against injury and accidents, and to enable them to perform the offices of regeneration properly: then again they are covered round with a kind of flesh in considerable quantities; and this again is enveloped in a coat or coats, as we see in various fruits, as grapes, berries and the like. We reason then from analogy to the present case, being supported also by the nature of the seed as made known by chemical experiments, and as it is disclosed by the microscope. And the inference is, that the seed is no simple substance, but doubly and trebly compound; and that each integral part of it is a kind of fruit, in the innermost parts of which the genuine sperms are laid up; which sperms are further surrounded by a common tunic, and these again as new compositions or combined forms, have stretched around them a tegument or sheath of the most general kind: the parts of the seed thus growing through three dimensions, before they are perfect and as it were ripe. The first dimension is perfected in the epididymides and vasa deferentia; the second and third, which is superinduced,*

in the vesiculæ seminales. Thus the several parts of the seed are as little eggs, surrounded with membranous envelopes and with an exquisitely fine shell, in whose innermost parts the true rudiments and stamina that are to supply the offspring with life and new generation are laid up, like seeds in fruits. And for this reason the female organs, of reception, conception and formation, are expressly suited to take to pieces again and let out this humor in the same order in which it was put together or fashioned in the organs of the men. These are the preliminaries which I have felt called upon to explain briefly, to enable the reader to comprehend the entire process of generation at a single general glance, so that afterwards he may proceed from this point to the survey of the several parts. And I hope to be pardoned if in the course of what follows, I make use of some common terms; and call for instance the first parts primitive globules; the other or second again compound globules, and the third, true compound globules: although it is plain enough, that these first essences of the seed ought not to be called either parts or globules; yet as we have to express them by intelligible words, such terms must be used as they will agree with analogically.

52. *With regard to the vesiculæ seminales themselves, they are the parts that collect and reserve the first essence of the seed or its simple globules, in which lies the true seminal and generative power; in virtue of which office the vesiculæ are the proper granaries and storerooms of the seed. Then again it is the vesiculæ seminales that act upon those really prolific, vital and seminal globules collected within them; namely, that digest them, roll them about, mingle them well with the grosser essences, and take care that they are duly made up into forms suitable for use, that they are perfected, and carried to maturity. In these respects the vesiculæ seminales are to be denominated Laboratories. Lastly, it is the vesiculæ seminales that excite the other genital organs to action; nay, that erect the urethra and its parts, as well as the corpora cavernosa and the penis; and thus throw forth their treasures, well kept and duly elaborated. In this point of view they are Dispensatories and Erectories. For they are placed midway between the urethra and the testicles, and look towards both; hence when we inquire for the proximate cause of the operations which occur in the process of generation, we find it in the vesiculæ*

seminales. We consider them therefore as the seat, the native land, the Island and the Paphos of Venus, where her little boy Cupid is born, and from which he aims his arrows, when any two breasts are said to be wounded. For the state of the whole case, and all the performances of venery, proceed from the state and changes of these vesicles; which therefore we must examine not slightly but thoroughly. Let us then now weigh one by one with a finer curiosity the several details of function just enumerated.

53. *The vesiculæ seminales receive, collect and keep the first essence of the seed, or its simple forms, or if the reader pleases, globules. This is manifest from the insertion of the vasa deferentia, which is such that little by little and successively they derive through the narrowest orifices all the prime seminal material from the testicles; after deriving, receive it; after receiving, collect it; and when collected, keep it for a considerable time. The anatomical connexion of the parts, when fairly examined, suffices to shew that this is the case.* Nothing can be plainer than that the testicles determine the whole of the substance that is extracted in the labyrinthine structure, into the vesiculæ seminales alone: or than that they receive it thus determined, and long retain and keep it when received: for as indicated above, the testicles derive none but its superfluous portion; and this successively: the mass increasing and enlarging the while from this successive increment. *Thus a quantity equal to generation is brought together; and thus the vesiculæ are not only repositories but also reservatories, for they consist of distinct cells, and may be opened out and unfolded like the intestines; moreover they are rugose or wrinkled, which allows room for their enlargement, and enables them to store away a lesser or greater quantity.* Each vas deferens flows into the middle portion of the corresponding vesicula, and this middle region is furnished with particular compartments and cells which can be opened out, as is shewn not only by its rugous character, and by the fact that it may be unfolded, but also by the effects of the case; for the seed may be kept in it for a considerable period, consequently may be collected there. This follows from the insinuation of the semen from the testicles through the vasa deferentia, which takes place successively and little by little; and also from the length of time that it is kept in the vesiculæ. *For in order that the semen may serve its pur-*

pose for the new generation, not only must the matter of quality be attended to, but also the sufficiency of quantity: now without the vesiculæ seminales this quantity could not be had. Were the semen drawn in immediately from the testicles through the vasa deferentia, and transferred through the urethra, into the womb, it would not produce an effect answerable to nature's laws. For instance, were the quantity scant, the effect would be rendered uncertain, and the most part of the seed would be dissipated and volatilized, perhaps even in the trabecular structures of the urethra and in the corpora cavernosa, or else in the blood and fibres, before the effect was reached; the same thing might happen in the womb. Quantity therefore is a desideratum in order to ensure the effect, and that the seed may not only line the parietes of the uterus, but also go up into the Fallopian tubes and the ovaries; which could not be the case if only a drop or two got in, and sticking to a portion of the surface of the uterus, were arrested there until its more intimate and volatile essence was dissipated. It is therefore necessary that the seed should crowd the doors that lead to the ovaries; furthermore that it should run through the expanded urethra; and on the way fill in and fertilize all the adjacent parts. Moreover, were it supplied in quantity from the testicles immediately, before it reached the middle of the urethra it would be scattered to the winds; this however belongs to the disquisition on the matter of quality.

54. *The vesiculæ seminales digest the simple and primitive globules of the seed derived from the testicles, roll them about, and mix them up with the grosser essences. This is plain from their very make, from their motive powers, in short, from the fact itself. Thus these vesiculæ not only give quantity, but also produce, and induce upon the seed, quality also.* The quantity is of no great import unless the quality be there; for it is the latter that gives the real power of prolification: a quantity of the right quality does the business. *The quality consists in this, that the quantity of prolific seed or primitive and truly vital substance be at hand and in place, but this bare substance, not well disposed into the proper form, because it is volatile and easily resolvable, is of no use, unless like the seeds of the vegetable kingdom it be again and again enveloped with particular membranes; and with these it cannot be enveloped, unless suitable connecting*

substances be at hand, to tie together as it were into bunches these primitive globules, and when so grouped, to wrap them up in a general tunic, and by repeated processes of composition to reduce them into answerable forms. Namely, according to the description of the seed, whose innermost parts and centres are occupied by the purest substances, while the grosser constitute the surface, just as in the little seeds of the vegetable kingdom: with this difference however, that animal seeds require to be most thinly and delicately coated, because they are to be again resolved and opened in the genital soil of the mother's womb, whereas the seeds of vegetables have to undergo the process in the bosom of the earth, and to this end their coats are comparatively thick, and only to be resolved by prolonged maceration and a species of putrefaction. *In order then that an abundance of such fluids, suitable for the formation and composition of the tunics, may run into the vesiculæ seminales, liquor of divers kinds and divers species is derived into those vesicles; that is to say, there runs from the arteries sent hither from the rectum and bladder a quantity of subtly sulphurous and urinous elements; from the sebaceous glands which beset the inner surface, a suitable and unctuous matter; from the cellular tissue of the peritonæum and bladder, liquids which are adapted for uniting these menstrua; lastly, from the fibres that come through the vasa deferentia, and end in the common tunic and the septa of the cells, and also from the proper nerves of the vesicles, that which is termed the animal spirit. Thus there is nothing absent that can be of service for uniting the seminal globules, and weaving the inward tunics and forms of the seed.* Anatomy shews us that numerous arteries are derived hither from the rectum and the whole extent of the bladder. The arteries make their approach to this threshold in the first instance, this being their quiet terminus and haven; and on the way they deposit all those refuse parts that are too gross, and would not be serviceable for the above use; letting drop this excrementitious portion both in the bladder and the rectum, and carrying none but the purer parts of the blood and serum to the vesicles. This is susceptible of particular demonstration. Moreover anatomy teaches us the mode in which the arteries are inserted into the vesicles. But glands resembling those of the sebaceous order, also thrown in a fatty, unctuous and viscid humor. This follows too from the mere existence of the glands, which the best anatomists acknowledge, and whereof they have discovered numbers on the internal surface of the vesicles. A certain liquid also is poured on by the cellular tissue, as appears not only from the continuity of its tunic from the peritonæum, the bladder, and the vasa deferentia, but also from anatomical experience; for when the vesiculæ are inflated or injected, the air or injection not only expands the vesicles of that side, but also passes into the adjacent parts of the coat of the bladder, and raises them, although it does not penetrate into the bladder. See Winslow and the other anatomists. This water too unloads its more feculunt serosity, as does the blood itself, in the bladder and elsewhere, and carries only the pure serosity to the vesiculæ. The animal spirit also is copiously sprinkled on, nervous fibre being continued not only from the testicles and epididymides through the vasa deferentia to the universal tunic of the vesiculæ and their septa, where it terminates, but also from the pelvic plexus. And whatever the fibres from these two sources carry, can be discharged nowhere but at the end of the fibres; but the charge of the fibres is the animal spirit, and that, in very considerable quantities. Thus nothing can be wanting to the ulterior composition of the seed; for whatever has power and faculty of copulating or combining, is ready at hand; thus we have here an agent spirit, a serum full of sulphureous elements, and at the same time another serum of a glutinous and subtly fatty character. This however is not the place to consider the mode in which the several elements are copulated. *But in order that these several elements may be properly combined, and each obtain their own places in a fitting manner, they require to be perpetually rolled round and round, and mingled together; and to produce this result, a gyratory motion is requisite, which nowhere in the body exists in greater perfection, or arises from more numerous causes, than in these little vesiculæ seminales. An intimate motion of all the fluid parts here is induced by the influx of the seminal globules from the vasa deferentia, and by the spirits themselves that flow in through the nerves; also by the other menstrua that are insinuated each through their own orifices. Moreover also by the fibres themselves of two-fold origin, particularly by the simpler fibres, which aspire to a like motion with the spirits they enclose, and by their common tremble renew that intestine motion in the way of instrumental causes. For be it observed that in the very affection and act of venery there lies a tremble and quiver of the purest kind, which is communicated to the nervous fibres throughout the system. But besides this intimate and intestine motion, another of a more general kind is put in by the muscular membrane, which covers and girdles the vesicles, and which is described as a robust structure by numerous anatomists. A motion still more general, nay, the most general, is given by the peritonæum and the muscles of the abdomen, for the peritonæum constitutes the outermost tunic of the vesiculæ, and communicates to them all the motion of the abdomen, and this motion is perpetual, and various according to all the actions, natural and voluntary, of the body and the lungs. Thus are there causes of motion here given, particular, general, and most general, all of which excite and impel the vesicles, and the fluids contained in the vesicles, into a gyratory or circulatory motion running from one cell to another, in a circle of motions. This gyratory motion follows the form of fluxion of the parts, which gives to the vesicles such a figure that they flow around agreeably to their nature; whence it may be inferred, that the parts of the seed are raised to their third dimension.* The purest parts of the seed, which flow in through the vasa deferentia, and carry an intestine motion into the whole mass, are the most active and mobile of all; so much so, that their motion is parallel with the form of the modifications in the purer aura, being vortical and perpetually gyratory. Similar also is the motion of the fibres, that renew the above motion; on which subject see our Doctrine of Forms and Modifications. Every time that aught of this sperm is instilled through the vasa deferentia, a motion of the kind is given; and at the same moment, by the fibres, which are everywhere scattered about through the septa, and thus brought to bear: moreover also, by the muscular tunic, which is excited every time that the fibre of the urethra, penis, and other genital organs is called in play. The peritonæum is constricted and expanded every time any natural actions exist, as in all respiration, all throwing and extension of the loins and arms; in struggling, personal contests, walking, running, riding, discharging the fæces and urine. But as this breathing motion is most general, it does not act on the innermost parts, but in a general manner kneads, presses, and rolls this seminal mass. From this triple motion, and also from the form of motion relatively to the form of the ducts, and of the consequent circulation through the cells, it may be inferred, that the genuine sperm is raised and compounded to the third dimension. The rules of the induction are proved by the rules of the doctrine of forms. *Thus the seed is perfected, and carried to maturity, that is to say, each part of it is carried to the third dimension, and gets its proper embodiment, to give it consistence, and ensure its due conveyance through the canal of the urethra into the womb. This maturation or ripening is produced not only by the action of time, but also by the continual circumvolution and kneading of the fluids of different kinds which are abundantly supplied. In these respects animal nature resembles vegetable, in that seeds require to be ripened before they can perform their use. For this end moreover, ripe and virile age is requisite. But in earlier life, before the more virile years commence, the testicles and vasa deferentia pour in only the vital sperms; the next period pours on a greater abundance of juice and copulating or combining moisture: thus the seed is perfected, and is emitted the more fitly for generation without the loss of its powers.* These particulars are intelligible by comparison with the seeds of the vegetable kingdom. They are also confirmed by the facts of the case, and particularly by the actual variety of the seed in color, consistence, and other respects. *The semen then gets its quality, which is very various, in the vesiculæ seminales; the like quality not obtaining in any two subjects. It may contain either little or much of vital or procreative sperm; also more or less of humor apparently seminal; or the sperms themselves may be more or less closely implicated and stuck in the parts of the humor; or they may be loosely coherent with it; and some of these circumstances may render their resolution in the uterus more difficult; or they may be of another consistence again, and grouped in another way. These varieties are indeed conspicuous to sight, in the color, density, quantity, chemistry of the object; for sometimes little insects or living corpuscles are represented in the seed. The cause of the varieties in quality depends on the time of life, on the degree of ripening, the frequency of venery, the degree and state of venereal desire, the constitution of the body, and the contingency of several diseases. The semen in the vesicles is first changed, infected and decayed into the matter of gonorrhœa; while the sperm in the testicles, epididymides and vasa deferentia is altogether uninjured. Thus all the quantity and quality of the seed is to be regarded as due to these vesicles, which on this account merit to be termed the grottos of Venus and Cupid: the crypts of love and desire.*

55. *But the vesiculæ seminales are also the prime and principal causes that by their own way excite the active desire and kindle the flame of love; for they are extremely sensitive, and convey the very state of the seed, and of its activity particularly, to the cerebellum, which straightway excited by the soothing, calls into desire all its fibres and the whole of its nervous system. For there are in the vesicles fibres and septa of the most delicate kind, and moreover a papillary or villous tunic, which as organs of touch, perceive every vibratory touch and modification; by their notice whereof the fibres of the cerebrum, which all over the body are implicated with those of the cerebellum, are roused up; hence too so are its sensoria; and desire and pruritus are the result.** This villous or papillary tunic is mentioned by Winslow. It covers the inside of the vesiculæ, and is the more sensitive in proportion as all corners of it are filled, particularly when a plentiful collection of seed stretches out the coats and papillæ, and intensifies the feeling. For this end nervous fibres are derived thither from the pelvic plexus. Sense can no more be denied to these papillæ than to the fibres of the universal skin, particularly when the fact itself proves its existence. Desire is kindled into pruriency when the quantity of the seed is large, or its quality too acrid and pricking. The circulation and intestine motion in the vesiculæ cannot fail to rouse the fibres, as well as the septa, which are of a membranous character. But these vesicles appear to have such a faculty of sensation put into them, that their papillæ actually feel the intestine motion of the genuine sperm, but the membranes the more general motion: hence the

* Let the reader notice here and elsewhere how completely Swedenborg makes use of the most advanced views on the subject of reflex action, as it is called; applying them however more catholically and easily to the explanation of the works of the body, than their modern advocates.—*Tr.*

variety of the affection or effect.* *The vesiculæ seminales are excited through another channel, namely, by way of the external senses; by objects seen, touched, kissed, embraced; nay, immediately by the very imagination of the like: whence the mind falls prone into the affection of love, and in this state the fibres both of the cerebellum and the cerebrum, consequently the fibres of the whole body, deliciously and intimately tremble, opening and unloosing themselves; and the same is the case with the internal sensory organs or the cortical glands. The effect strikes at once to the vesiculæ seminales, where the very matter of love is heaped up, which is at once excited to a similar tremble and motion. Thus desire is raised, and kindled into lust; for now the feeling returns from the vesicles through the fibres to the cerebrum and cerebellum. Thus the vesiculæ seminales are both the first and last bounds from which the venereal affection proceeds.* It will be shewn below that the testicles themselves are not such bounds, unless the person be a man of pure affection or genuine conjugal love.

56. *The vesiculæ seminales are also the first impulsive causes that erect the urethra, and the corpus cavernosum, that is to say, the penis, which surrounds it, for by exciting the motion of the seminal substance they enclose, they rouse the beginnings of the urethra, its caruncula and the other parts, which vibrating with this soothing or titillation, makes the several pores open, for the nervous spirit and the blood to flow into without impediment; which effect is a continuation of the cause. For as the least fibres are set open by that desire, so also are the compound fibres by a similar force springing up in the vesiculæ. Thus the vesiculæ seminales are not only repositories but erectories.* We shall treat of the proximate causes of erection in the following pages. But of the proximate cause, the proximate or efficient cause is the excitation of the seminal humor. The tunics and septa of the vesicles are continued through the two ducts all the way to the caruncula, where the accelerator muscles lie; and all the vibration that is impressed on the little tunics and septa of the vesicles passes immediately to the caruncula, whither those orifices penetrate. But of this more anon.

57. *The causes however of the excitation of the seed of the*

* Let the reader continue to observe the analytic penetration of Swedenborg's use of the doctrine of reflex actions.—*Tr.*

vesicles into that intestine motion that rouses the fibres of the urethra, are many in number, that is to say, both internal and external. The intimate or inmost exist when the cerebrum and cerebellum, roused up by objects either external or internal, act through the fibres on the vesiculæ seminales and their spermatic humor. The internal or middle causes exist when the humor itself abounds, or is of that quality that on the least excitement it rushes into a motion and pruritus of the kind. The exterior causes again are when the penis, the glans penis or the parts continuous therewith are rubbed; in which case a similar rubbing flows by continuity all the way to the vesicles, and the animus now itself concurs, being similarly chafed. This follows from what we have said above, for given the cause and the effect is also given. *The vesicles and not the testes are the parts that produce the erection and expansion of the urethra. This is abundantly clear from the case of those who have no testicles, or where these organs have been excised,—from castrated persons who have lost them from whatever cause, and who nevertheless during erection of the penis emit a humor not unlike the seminal fluid. It is also shewn by the stiffening of the penis in sleep, when either the fibres of the cerebellum are quite detached from those of the cerebrum, and left to themselves, or when the urinary bladder is distended, and by consequence the vesiculæ seminales are increasingly expanded in all directions; and the seed, compressed in the cells, makes a strong push for its orifices, viz., the beginning of the urethra and the caruncula.* If the testes took rank as the efficient cause of erection, the penis of eunuchs could not possibly swell up; nor could this happen unless something like seed were contained in the vesicles: but as this is made up of the supply of humor treated of above, it is a sign that the efficient cause of erection lies in these vesiculæ and their inciting humor, and not in the testicles.

58. *Furthermore the vesiculæ seminales dispose not only the fleshy appendage of the penis and excite it to coöperation, but also the vasa deferentia and the testicles; which shews that the vesiculæ act from a central point, upon both the parts that precede and those that follow them. For the proximate coat that encircles the vesicula, and moreover the septa of the cells, are continuous with the fibres of the vas deferens, and hence with the fibres of the epididymis and of the tunica albuginea testis. Still it cannot pass into the compages of the testicles unless the intes-*

tine motion be that of the genuine seed, or unless it consist of the tremble of the true simple fibres. This follows from the connexion of the parts; for the fibre of the testicle is continued into the vesicles; but the course is not from the vesicles into the testes. All feeling that gives birth to action is propagated in the course of the antecedents of the fibres, or from last to first; for example, from the organs of touch, taste, smell, hearing, to the cerebrum and the beginnings of the fibres: on the other hand, motion is propagated to the consequents, or from the beginnings to the ultimate ends, *e.g.*, from the cerebrum to the muscles. The like is the case here in the vesicles, the origin of whose fibres is in the testes; wherefore the so-exquisite sense of the vesicles goes back to its principles in those two parts. Still the latter are not therefore excited to action; for the efficient cause of the motion of the testes arises from the beginning of its fibres or from the cerebellum; they themselves not being the cause of their own motion. This is a wise institute for preventing the testes from breaking out into action every time that the vesiculæ seminales do so; for were this to be the case, they would wheedle from the blood and fibres an excessive quantity of spirits, and exhaust the body; for which reason they are not excited more strongly than the desire itself kindles the animal imagination.

59. *While we are on this topic, and before going further, it may be proper to add, that there are as many degrees of sexual love as there are degrees of composition in the parts of the seed, or as there are degrees of composition in the fibres themselves. The first degree is the purest, or the conjugal degree, and is properly to be entitled Love. This degree respects no end but propagation, and is the innermost of all, that of the very soul, and which never regards the effect without at the same time regarding the end in the effect. It excites the simple fibres, consequently the testicles themselves immediately, and by these means the vesicles, or what amounts to the same thing, it excites that purest vital essence that constitutes the genuine sperm. This love refreshes and renovates the nervous system, but never destroys it; it vivifies the very inmost factories of the cerebrum and cerebellum, and excites them to restore an equal portion of sperm to that which they spend. The second degree of sexual love is that which is excited not rationally but naturally; namely, by objects appealing*

through the external senses by touch, kisses, embraces, conversation, fancies; and it has no end but in so far as it is seized with desire, and burning to enjoy the delights insinuated by sense and exalted by the imagination. This does not excite the simple fibres, but the fibres of the first composition,—not the first vital essence, but the spirit thereof or the animal spirit. It does not immediately invade the testes, but the vesiculæ seminales, from which it is reflected to the testes; consequently not the first but the second composition of the seed, to the dimension whereof this degree of tremble is adequate. If this degree of venery occurs too frequently, and spends more than the superfluous part of the seed, it takes life from the nervous and sanguineous systems, producing lassitude, heat, cold, and other conditions, which are the cause of diseases arising from deficiency of the spirits. But the third or last and lowest degree of venery is lasciviousness, such as is excited beyond what nature requires, by divers external causes, as friction, love-potions, drugs, and other unlawful means of the sort. This is to the utmost degree destructive to life, and attacks none but the grosser fibres, the blood itself, and the last composition of the seed, which irritated to motion, infests and in a short time destroys the organs. It begins in the last sphere, that is to say, in the blood or arteries, and has no end but lasciviousness; but all this without any previous desire.

60. *Meanwhile as the seed of the vesiculæ seminales occasionally has too great a viscidity by reason of its retention in those organs, and is sometimes rendered too slow and too little fluid to allow of being vivified and excited to the proper vital motion, in the act of emission, so a further means of efflux is provided from the vasa deferentia immediately into the urethra: and thus the seed is so attempered and qualified that it bursts forth with fuller effect and surer use.* There is also a way from the vasa deferentia into the urethra, but the seed which traverses this meets first with that of the vesiculæ. All cases and dangers are met; in order, namely, that there may be no natural failing to mar the attainment of the end, which is the propagation of offspring. The sperm of the vasa deferentia is the vital and prolific thing which in this case is mingled in the very act with the seminal material.

CHAPTER VI.

THE PROSTATE GLAND.

61. HEISTER. "The prostates, or more properly speaking, the prostate, for it is a single body, is a globose and heart-shaped gland, situated just in front of the neck of the bladder, and completely surrounding the beginning of the urethra. . . . It is about the size of a walnut, and has two natiform prominences on its back part. It also has an eminence, the caput gallinaginis, in the urethra, and two little orifices, which are common to the ejaculatory ducts and vesiculæ seminales; and a . . sinus is frequently found between these orifices. The prostate is of a glandular and spongy texture, and is surrounded by a strong membrane. The minute foramina or excretory ducts of this gland excrete a whitish humor from its little crypts into the urethra. In the human subject there are ten or twelve of these foramina; in dogs they are very numerous. The prostate has its vessels in common with the vesiculæ seminales." *(Comp. Anat.*, n. 228.)

62. WINSLOW. "The first portion of the urethra, or that which is not covered by the cavernous substance, and which from the bladder to the bulb is only a simple membranous canal, is sustained by a large, solid, whitish mass, of the shape of a chestnut, and situated between the bladder and the bulb of the urethra, its base being toward the bladder, the point toward the bulb, and the sides looking upward and downward. This body is termed the prostate, . . . in the plural number, because it appears to be divided into two lobes, by a hollow groove, which runs along its upper side from the base to the apex. The first portion of the urethra lies in this groove, adhering very closely to the mass of the prostates. The body of the prostates lies on

the rectum, and the apex is under the internal lip of the arch of the ossa pubis. Their substance is spongy, but very compact. In each lobe there are several follicles, which open into the first portion of the urethra, toward the bottom of the groove. . . . *(Exp. Anat., Tr. du Bas Vent.,* n. 534—536.) At the bottom of the cavity of the first portion of the urethra, or that which lies within the prostates, there is a small oblong oval eminence, large on the back part, and terminating forwards in a point: this eminence is the caruncula, caput galli, or verumontanum. The large portion of it is commonly perforated by two holes, sometimes only by one, seldom by three; these are the excretory orifices of the vesiculæ seminales. . . . Each orifice has a small thin membranous border, which may serve as a valve to the excretory ducts of the vesiculæ. On each side of the large portion of the caruncula, there are five or six holes ranged in a crescent round its lateral parts; these are the orifices of the excretory ducts of the prostates, which ducts come from the follicles, . . . and run in an oblique course to the orifices, in a kind of membranous duplicature. *(Ibid.,* n. 543, 544.) The superior prostatic muscles are two thin planes attached to the upper part of the inside of the small rami of the ossa pubis, from whence they spread over and are inserted into the prostates. The prostatici inferiores are small transverse planes, each of which is attached to the symphysis, between the ramus of the os pubis and os ischium, and from thence runs transversely, till it meets its fellow from the other side under the prostates, to which they are both strongly connected, serving as a girth to support these glands. They may be considered as the small or internal transversales, and the other two may be distinguished as the great or external. They have likewise some connexions with the point at which all the other muscles of the urethra and penis meet." *(Ibid.,* n. 574, 575.)

63. Boerhaave. "In the same place where the outlet of the vesiculæ seminales opens into the urethra, the prostate surrounds the latter. It is a single, continuous, conical gland, environed with muscular fibres. It is made up of twelve distinct glandular groups; and the glands of each group terminate by their excretory ducts in a single little sac, into which they pour their humor. Hence we have twelve distinct sacs, opening by

as many distinct and considerable ducts into the cavity of the urethra, in such a manner that the orifices surround on all sides the outlet of the vesiculæ seminales. Thus it is that at this spot the seed and the humor of the prostate are accurately mixed: the vesicles and prostates being moreover surrounded by the same muscular membrane. The humor made in the prostate is bland, fat, white, and abundant." (*Inst. Med.*, n. 652, 653.)

64. MORGAGNI. "The humor that is secreted in the prostate gland, is sometimes so much altered, that in some bodies I have found it coagulated into a number of granules of a dirty yellow color, seeming like tobacco; and these I observed not only in the urethra at the orifices of the ducts of the prostate, but they pervaded almost the entire substance of the latter, wherever it was cut into." (*Advers. Anat.* IV., anim. 14.) In several subjects also Morgagni found in the caruncula a sinus that presumably communicated with the seminal ducts, for when the vesiculæ seminales were compressed, the semen came out of it. (Anim. 3.)

65. In *order to enable the vesiculæ seminales to act upon the urethra and the muscular flesh surrounding it, for receiving and expelling the seminal humor, and for disposing all things to the effect of the end, they, the vesiculæ, have need of the prostate gland, which must be close to them, and in contact with them; and must engird the orifice or beginning of the urethra. Moreover they have need of those two general* bodies termed the verumentanum, caput galli, or caruncula. For instruments and organs are necessary in order for the principal cause to extend its sphere of operation. The instrumental causes comprize all those parts that are visible in the urethra and about the urethra, the chief of which however is the prostate, which besides serving in the function of mediation, also strengthens the beginning of the urethra, and disposes it for proper action. And this all the more, inasmuch as innumerable varieties occur in the dispensation of the seed, which in order to their attainment, require the organs to be disposed to correspondence in all sorts of ways.* The vesiculæ seminales and the prostates communicate very closely, as is evident from their mutual nearness, contact, and insertion. Furthermore, what is termed the muscular coat is common to them both. Then again the emissary ducts of the vesiculæ and the prostate meet in one raised spot, viz., the crista galli. *Anatomy also shews us, that the prostate completely surrounds the beginning of the urethra, leaving only a fissure, which affords a passage to the two excretory ducts of the vesiculæ seminales. And an almost similar fissure is seen at the end of the urethra, these two fissures communicating with each other. For the urethra,* i. e., *its membranous cavity, is continued from the prostate; and in this way*

* The author uses the term *general* in his technical sense; indeed, in the sense opposed to *particular*, but not applied abstractedly, but organically.—*Tr.*

the action of the urethra depends on the action of the prostate, or on that of the vesiculæ seminales by means of the prostate. But now as the vesiculæ seminales act in the first instance upon the urethra, and as this is also the case with the prostate, it follows that there is some middle or third uniting body, which combines the actions of both: this third body is the caput galli or caruncula, to which as a common terminus flow in two ducts from the vesiculæ seminales, and several from the prostate; shewing that the action of both commences from it. This caruncula is clearly a kind of appendage, and third body, or means of union on both hands, as appears from the insertion of the ducts of the vesiculæ seminales, and of the ducts of the prostate, which however enter it in different ways. *The prostate also keeps watch to prevent anything from the vesiculæ from coming forth into the urethra, before the ultimate effect is gained. The caruncula is a similar check. Were it not for these parts, and the assistance they bring, the seed would drip away little by little, in scattered quantities, before the ultimate effect alluded to, that is to say, before the act itself, and the end in view would be lost.* These parts then are to be regarded as two bars, that carefully prevent the seed from being thrown forth before it has acquired the due conditions of quantity and quality. This is evident not only from the organism, but from the fact also. Unless these shields or bars be broken violently through by some particular force, no passage is afforded, but the knocking at the door is fruitless: the violence however succeeds so soon as they can no longer make a resistance to the impelling force. As to the fact, however, there is no room for doubt. In this place it comes to be deduced only partially, because from anatomy: but we must moreover explore the mode itself from the ground of the organic conformation.

66. *The causes of the excitation and propulsion are internal and external. The instrumental causes of these are the vesiculæ seminales, the prostate and the caruncula. The internal causes are all those that excite the desire of venery. They are then first properly internal, when they beset the halls of the cerebrum and cerebellum, or animal mind. As soon as the cerebellum or its little sensories are beset with this desire, then all its fibre from the beginning to the ends is not only opened out, but also quivers*

and trembles in the purest manner; and the consequence is, not only a discharge of the spirits, but a similar tremble or internal motion of the seed, particularly in the vesiculæ seminales, where they (the spirits) reside in gathered store. This motion rouses up the vesicles, their walls and partitions, and sets them in a constant effort to rush out of doors; and this effort is all the stronger the more seed is present, much as is the case with respect to the other evacuations, but in the latter more grossly, bluntly, and sensibly; as happens in the discharge of the urine from the bladder and of the fæces from the rectum. This it is which constitutes pruriency, or the itch for coition: the very feeling and sensible effort at the time prove what we are saying. The case is the same if the quantity of the seed be excessive; or if it be less, and still the intestine activity urges it forth, which may happen from different causes. But of these we treated above. Such then are the internal causes.

67. *The causes of the excitation and propulsion of the seed are external also: these spring from causes in the urethra as well as without the urethra, and stimulate to the sexual act. The gland of the urethra, or the penis, being papillary, is very sensitive, and when softly stroked, by contact with the vulva and by the repeated motion of the prepuce upon it to and fro, it produces a similar but grosser effect. So too at the same time the band under the urethra, which is continued externally around the scrotum and perinæum all the way to the anus, and internally by the urethra all the way to the prostate; and again, the middle partition of the scrotum, produce the same effect.* These causes we term external, because they arise immediately from the fountain-head and beginning of the fibres, or from the ends of the same, and from the last terminus of the organs. Every papilla is sensitive, and conveys its sense backwards, or to the antecedents of its fibres. The sensation is a touch, and the touch may be likened to a tremble of that degree; which is too pure to be felt as tremble, but is perceived as a modification. *The whole of this tremble or modification meets in the vesiculæ seminales and prostate gland. The modification that arises from the internal cause is concentrated in the vesiculæ seminales; that which arises from external causes is concentrated in the prostate gland: both lastly in the caruncula,*

which is the means of union between both. Thus the prostate receives the internal cause mediately by the vesiculæ seminales, and the vesiculæ, the external causes mediately by the prostate. The connexion of the fibres proves the truth of this. For all the fibres that are sent by the cerebrum towards the genital organs, are in the first place transmitted to the testicles, from whence they are reflected through the vasa deferentia to the vesiculæ seminales; besides which, fibres from the pelvic plexus are also carried thither: thus of course the vesiculæ receive the last impetus of the modifications of the cerebellum. The fibres however that are put forth to the urethra, and through the urethra to the glans penis, are continued from the prostate; hence the last terminus of that modification, or rather friction, is found in the prostate; in other words, the external causes there end. So it is that the prostate communicates its tremble to the vesicles; and these, theirs, in the prostate; and both of them in the caruncula. *Before the venereal desire, from whatever cause arising, can burst forth into act, the internal and external causes must necessarily concur. The inward cause alone does not produce the effect. Nature, in proceeding from her innermost causes to her outermost causes, namely effects, goes by the scale of degrees, that when she arrives at the effect, those that we have called external causes may be called into play there again. For no effect in the ultimate sphere is produced by the innermost cause, without an orderly progression of means or middle causes.* When we consider the act of coition or venery, we find plainly enough that external causes concur to the whole effect. For during that time the gland of the urethra, viz., the penis, is in the highest degree of expansion, hence in the highest of sensibility: and during the copulation and congress this gland is touched in divers ways, and stroked by what it meets: the prepuce too is likewise drawn over it backwards and forwards: the frænum, or urethra, is alternately tightened and let go: so there are external causes at hand which conspire with the internal to one and the same effect. *Until all the causes meet and are raised to the highest power, the effect is not produced; meantime the seed of the vesicles acts on its ducts, and the prostate likewise on its: the caruncula however is not opened to transmit these humors until the last force has come on; which is so violent, that the caruncula is opened not spontaneously but by the force applied, and makes room for a passage. Thus there is no gentle transmission but an ejaculation of the seed.* These ducts are of very small calibre, and do not transmit the least dew of moisture unless force be put in. Thus the two ducts of the vesiculæ seminales pass through the naked base of the caruncula, while the ten or twelve ducts of the prostate, fine like the others, decussate with each other, and in a manner close their own passages by their mechanism. The ducts of the vesiculæ seminales form in the caruncula a species of valve; see Winslow. And moreover the vesiculæ are corrugated, and may be distended to a great extent, and by their swelling afford room for the urgent seed before it bursts forth. The prostate also has its appendages in the prominentiæ natiformes, and possesses in fact peculiar receptacles apart from its follicles, which are many in number. The caruncula too in some subjects is provided with a little sinus of its own. All which shews that every precaution is taken to prevent the seed from being exiled and carried away to the ultimate effect, and bursting forth without the aid of the efficient cause. And moreover, if by long protracted continence a quantity of seed is collected, the vesiculæ seminales can send back their superfluity again into the blood, for there are veins close at hand to absorb it. And the glands conspicuous therein seem to be of an amphibious or hermaphrodite nature (like those that are visible in the intestines and the skin, of which we have spoken elsewhere); and by virtue of this double quality these glands can either transmit the genital humor into the vesiculæ, or can reabsorb it: for it is certain that this humor passes back often into the blood. But whether anything of the kind happens in the prostate, is a question: on this subject we may have our suspicions, but dare not say anything positive.

68. *To enable the seminal humor to burst forth, and break through its barriers, not only is a quickened intestine motion of the genuine sperm necessary, but also at the same time a similar correspondent motion of its ulterior compositions or subsequent compounds; but into the latter motion they cannot be excited by anything short of external causes. And still in addition there is required a reciprocal action of the tunics and septa, both those*

composing the vesiculæ, and those which enter into the structure of the prostate and its follicles. In the act itself or the last effect, the force of the muscular tunic is needed, or of the muscles that belong to the prostate, and which are the superiores and transversales: these, when called into play, superadd a force that overcomes the resistance: hence in the act there is a kind of convulsion of the muscular fibres, particularly those of the prostate. For the superior muscle, attached as it is to the os pubis, and the transversales attached between the os pubis and ischium, send moving fibres over the prostate, and thereby form a kind of muscular tunic: see Winslow and Boerhaave: this coat is also continued in a measure over the vesiculæ. Thus when these fibres act, they compress both the prostate and vesiculæ, and in the most powerful manner when coition is going on: that is to say, when the internal and external causes combine in their highest excitement. In the healthy natural state this force is so proportioned that the resistance is overcome, but only in the last degree of the impelling causes. *The muscular fibres of the prostate are not themselves excited to the highest or the convulsive degree of activity, unless moreover a cause be added which is not excited previously to the act itself, that is to say, in the reciprocal agitation of the os pubis and ischium; by which agitation the fibres, contracting and relaxing alternately, put the utmost force of action into the muscle. Were it not for this accessory cause of the action of the muscles, in the healthy state, when there is a natural resistance, no effect would follow, for this force it is that overcomes the resistance or reactive force.* I do not intend in this place to treat of the causes of the action of the muscles: but it may be mentioned, as we pass on, that these causes are internal and external. The internal spring from the principles* in both the brains; the external arise in the body itself, by frictions namely, and tremulous vibrations; as is observed in the phrenic nerve, which, when subjected to friction, excites the whole diaphragm to contraction. And touch applied

* The reader will not fail to notice that Swedenborg sinks causes, effects, principles, and other abstractions in substances, and never regards them long apart from substances. With him causes and principles are organic things. In the present passage, principles are cortical substances.—*Tr.*

to the muscular fibres in the heart and the stomach, even after the parts have been cut away from the body, affords the same phenomenon, as innumerable experiments testify. The like may be said of traction, as again proved by many trials; for when traction is applied to a nerve, the muscle is powerfully contracted. This obtains in the present case by the before-mentioned agitation of the person. *All fibre* expands and swells out under sweet and soothing friction; and the fibre in thus swelling, shortens or diminishes in extent: hence at this time the muscle is contracted, and acts on the space that it encloses. This is the reason that while the intestine motion of the fluids and the blandly-tremulous motion of the fibres is going on, the whole of the motive fibre around the prostate and around the vesiculæ seminales is contracted; and thus by virtue of internal and external causes, the fibre of these said organs all acts on the inclosed liquid, and urges its escape; not however still to such a degreee that the reaction is overcome until the local agitation of the bones of the pelvis comes in aid.* But these points will be explained more clearly further on, as well as in the chapter on the Muscular Fibre. *In order then that the act of coition may not be hindered by the acts of evacuating either the bladder or the bowels, particular provisions are made. Thus for instance the prostate lies on the intestine, and the intestine is so connected to the bladder that the two cannot act simultaneously; for when the rectum swells up, as it does during evacuation, the prostate is pressed down, but not the vesiculæ seminales: and at this time all the moving fibres of the muscles are relaxed, so that there is no action of the tunics or of the individual fibres upon the prostate and vesiculæ, but only a general action, whereby no libidinous excitement is produced. For this reason too we push the pelvic bones forward during evacuation, but in the act of venery the pelvis is allured backwards.*

69. *The humor which escapes in abundance from the prostate, which is indeed a gland, and which humor is sprinkled into the urethra, is exactly mingled, and this, close under the caruncula, with that spermatic humor that flows in from the vesicles, and with that pure seed that flows from the vasa deferentia imme-*

* We use the singular number, *fibre*, with the author, in a certain abstract or universal sense.—*Tr.*

diately into the urethra. The ducts of the vesiculæ seminales pass through the caruncula; likewise the ducts of the vasa deferentia, which combine with the former. The prostatic ducts, however, to the number of ten or twelve, pass through a particular membrane, and disposed about the sides of the caruncula, eructate their humor only close beneath it. Thus it follows as a matter of course, that these humors are accurately mingled: *and this, in such a fashion, that the humor of the prostate is scattered about over the whole circumference of the urethra which lies immediately under it; smears the entire wall of its cavity: and that the seed or genital humor is intermediate, and cannot leap against the walls until it has mixed itself in with the humor of the prostate.* The lateral ducts scatter their humor to the sides; the middle ducts, to the axis: and in this way the seed of course cannot possibly touch the walls excepting through or athwart this very humor. Nature appears to have had the same limitation in view in the obliquity and decussation of the prostatic ducts; for they run obliquely, and intersect each other: see the writers on anatomy. Hence when this humor is violently injected, it is carried in these directions, that is to say, round all the sides in circumgyration. *Wherefore this humor, intrinsically bland, milky, fat and abounding, (and not unlike the humor that is excreted from the arteries into the vesiculæ seminales,) at once absorbs, envelopes and enfolds the seed of the vesicles and vasa deferentia, and prevents any portion of it that is not yet sufficiently made up, formed and ripened, from impinging upon the walls, from being resolved back into its primitive and volatile essence, or from falling into the veins and little pits of the urethra, and thus perishing. On the other hand, it enables all the genuine seed, enfolded in this humor, to pass safely throughout the urethra, to the place of exit, and to bring out the destined use.* This humor is similar to that which is excreted by the arteries into the vesiculæ seminales; the arteries being the same for both organs, and arising from the same trunks: hence the humor is homogeneous, or suits the seed exactly. The sperm from the vasa deferentia, which immediately passes down into the urethra, is not yet full grown or entirely formed, but is virgin, tender, pure, volatile; now unless it were instantly enfolded in the humor of the prostate, at the least contact with the urethra or its walls, variously hollowed out as they are, and possibly also foul with the residue of the urine, it would at once be resolved and dissipated. There is also a similar sperm in considerable quantity in the vesiculæ seminales; for this which has recently come into them, cannot yet have been brought to its ripe state; and this likewise, unless so enveloped, would fly away. Whenever there is an intestine motion, solution and change of many parts is going on. Likewise if by considerable delay and prolonged rest the seed be too much condensed, it is at once irrigated or bedewed by this moisture, and conveniently dissolved. Thus this humor is not only the vehicle of the semen, but also its conservation and restoration. It also increases the quantity, being itself copious, and much exceeding the seminal quantity from the vesicles; so that this semen emitted from the vesiculæ, on meeting this humor, does not fall into an empty cavity. Were it not for this enfolding humor, the very small quantity of sperm from the vesiculæ seminales would be dissipated without effect even in the womb, and before ever it came in contact with the ovaries, would be absorbed in the lacunæ and veins of the womb. The afflux of this humor appears to be so great, particularly in the act, as to constitute it the main part of the seed; indeed, without it such repeated acts could not take place, for no more runs into the vesiculæ seminales than suffices for the supply of the sperm: the remainder is reabsorbed. In saying thus much I have endeavored to give an idea of the use of this humor. *If we properly consider the operation of the prostate together with the coöperation of the vesiculæ seminales, it may appear, that this humor of the prostate runs before the humor of the vesiculæ, in order like a bride to be ready in its bed to receive the bridegroom.* This follows from that connexion of the active forces of which we have already spoken. For the prostate has muscles of its own, and is compressed immediately by the muscular coat: not so the vesiculæ seminales, which appear to borrow their muscular coat from the prostate; hence it is everywhere more present and stronger on the latter, and also prior; namely, to prevent the prolific or fruitful seed from slipping away. So perhaps this humor also comes forth the first, to receive in the chamber the approaching seminal humor. Thus all the means that can possibly be required for the proper attainment of the end, are provided with the utmost wisdom. If there be a sinus in the caruncula, according to the observation of Morgagni, the seed tarries therein also before it is expelled. Sinuses however appear to be formed by interrupted coition, when the seed pushes forward into the ducts, but is not emitted: hence often gonorrhæa.

CHAPTER VII.

THE URETHRA.

70. HEISTER. "The urethra is a membranous, nearly cylindrical canal, continuous with the neck of the bladder, and reaching to the extremity of the glans penis. . . . It is situated in the groove formed on the lower side of the penis by the interstice between the two corpora cavernosa. It is not a straight tube, but is bent in a remarkable manner. It is twelve or thirteen inches long. Its calibre is about equal to that of a goose-quill. It is made up of two robust membranes, an inner and an outer, between which lies a spongy or cavernous substance, in which certain little glands are described by some anatomists. The bulb or protuberance of the urethra, is that part of it next the prostate gland, which is thicker than the rest, an inch long, and bears some resemblance to the bulb of a small onion: it is of a thickish spongy texture. On the inside the urethra exhibits various little foramina, some round and others oblong, and various little canals, from which in many cases a viscid liquid may be expressed which lubricates the urethra. . . . Cowper's glands, also called the mucous glands, are described as three in number by that anatomist. Of these, two, one on each side of the urethra, lie between the accelerator muscles and the bulb: they are of an oval figure, a little compressed, and about the size of small beans. They convey a clear mucous liquid into the urethra, some, by a particular excretory duct, nearly two finger-breadths long, and which perforates both the coats of the urethra. . . . In the most of subjects however in which I have sought for these glands, I have not been able to find them, at any rate not of the size that they are represented by Cowper:

and this leads me to doubt greatly whether they are always present. *(Comp. Anat.*, n. 229, 230.) Morgagni indeed affirms that he has found them in the most of subjects, and sometimes larger than a small kidney bean. *(Ibid.*, not. 29.) Cowper's third gland is a single corpuscule, which according to him is situated in the angle formed by the curvature of the urethra under the ossa pubis, within the spongy or cavernous body of the urethra. He figures it about the size of a lentil seed. But I have not yet been able to see it. The gland of Littre lies between the two membranes of the urethra, close in front of the prostates. It is of a dun red color, an inch broad, and two lines thick; it surrounds the inner membrane of the urethra like a girdle, and perforates it by a number of very minute mouths, through which a mucous liquid is supplied for smearing the urethra. The vessels of the penis, urethra, and of the above glands, are, 1. Arteries which come from the hypogastric and pudendal arteries. 2. Veins, provided with valves, which take back the blood to the veins of the same name: in the penis however, and under the ossa pubis, they first form surprising anastomoses called by Santorinus (p. 193) the labyrinth. The nerves come from the last nerves of the sacrum. The lymphatic vessels are finely delineated by Cowper and Drake." *(Ibid.*, n. 230, 231.)

71. WINSLOW. "The urethra is the third of the spongy tubes which make up the principal part of the penis. It adheres to the corpora cavernosa through the whole length of the inferior groove formed by their union. It differs from the other two, both in being narrower, and in forming a true canal. In substance it is spongy or cavernous, except a small portion next the bladder; its inner and outer surfaces, or convexity and concavity, are membranous. It is at first no more than a membranous canal continued from the anterior opening of the bladder, at the place called the neck of the bladder, which is a name that would be more proper for this first portion of the urethra. About a finger's-breadth and a half from its origin, it encounters a cavernous substance like that of the corpora cavernosa, only smaller, and which surrounds it through the whole extent of the inferior groove of the corpora cavernosa. But before this spongy substance begins to surround the urethra, it forms an oblong body

like a pear or onion, which is connected only to the lower convex side of the canal, and afterwards being split on each side, invests it quite round. This body is called the bulb of the urethra, being larger than any other part of that canal, and it is divided interiorly by a very fine membranous septum, into lateral parts; and therefore when it is inflated, it appears double or with two heads. (*Exp. Anat., Tr. du Bas Vent.*, n. 530—533.) The inside of the canal of the urethra is lined by a fine membrane, full of capillary bloodvessels; and its surface is perforated by a great number of oblong holes or small lacunæ of different sizes, the largest of which lie near the glands. These lacunæ are the orifices of the minute excretory ducts of the same number of small glands, which are dispersed throughout the substance of the urethra. The ducts run for some way in the spongy substance, along the convex side of the internal membrane of the urethra, and open obliquely into the great canal. The edges of the lacunæ are semilunar, or crescentic, because of the obliquity of their opening. A little way from the beginning of the cellular substance of the urethra, we meet with two lacunæ larger than the rest, and whose ducts are very long. These lacunæ and ducts lead to two glandular bodies, situated on the two convex sides of the spongy substance of the urethra near the bulb. Each of them is about the size of a cherry-stone, but they are oblong and flat, and covered round entirely by the accelerator muscles. . . . There is a third body of the same kind situated more anteriorly. The cavity of the urethra . . . is not everywhere round; towards the gland it becomes broader as well as flatter on each side, especially in the gland itself, where there is a kind of oval or navicular fossula. This canal terminates at the extremity of the glans by a narrow oblong orifice or fissure, which is much less than the rest of the cavity. . . . It seems to be surrounded by fleshy fibres. (*Ibid.*, n. 553—557.) The spongy substance of the urethra having reached the extremity of the corpora cavernosa, forms a large head called the glans, which is the common capital of the three spongy pillars; with this difference however, that it is a true continuation of the spongy substance of the urethra, and only adheres to the extremity of the corpora cavernosa without any direct communication. It is for this reason that if we inflate the spongy substance of the urethra, the gland is presently inflated as well as the bulb, but no air passes into the corpora cavernosa; but when we blow into one of these bodies, the air passes immediately into the other, the urethra and glans remaining as they were. The figure of the glans is that of a rounded cone, a little flattened at the lower part, and with an oblique prominent basis, the circumference of which is something greater than that of the corpora cavernosa. The spongy substance of the glans is thick and uniform next the corpora cavernosa, but next the urethra it is perforated by a continuation of that canal, and is there no thicker than the urethra before the formation of the glans. Thus the canal of the urethra does not lie in the middle of the glans, but continues its direct course through the lower flat side of it. . . . All the convexity of the gland is furnished with a fine villous substance; and this again by a fine membrane, resembling the red part of the lips. The circumference of the base of the gland has a double row of small papillæ, which may be regarded as sebaceous glands. . . . (*Ibid.*, n. 537—542.) The musculi-bulbo-cavernosi, commonly termed acceleratores, form first of all a penniform muscle, by means of a middle tendon, attached to the lower part of the interosseous ligament of the ossa pubis, . . . and to the union of the musculi transversales with the cutaneous sphincters of the anus. From thence they pass in an expanded form under the bulb of the urethra, covering the bulb and the urethra itself, and adhering in some measure to both, as high as the origin of the suspensory ligament, the middle tendon answering to the septum of the bulb. Afterwards the two fleshy planes separate, and run obliquely right and left from behind, forward, and from below, upward, embracing the two corpora cavernosa, in the sides of which respectively they are inserted. The middle tendon adheres very strongly to the lower part of the septum of the bulb, in which, and in the urethra itself, several of the fibres of these muscles are attached. The transverse muscles, called also triangulares, are two oblong, narrow, fleshy fasciculi, inserted each by one extremity in the root or beginning of the ramus of the os ischium; from whence they run transversely along the edge of the interosseous ligament of the ossa pubis, as far as the apex of the prostates, where their other extremities meet, and form commonly a kind of digastric muscle, the middle of which gives a common point of insertion to the muscles of the urethra, and to the cutaneous sphincters of the anus. (*Ibid.*, n. 571—573.) The arteries of these parts come principally from the internal iliac or hypogastric arteries, and some also from the external iliac and the crural. (*Ibid.*, n. 576.) The great middle vein is formed by the union of the hypogastric branches, which after passing on the two inner sides of the pelvis, meet about the middle of the arch of the ossa pubis. At this place we observe a venous plexus, which covers the upper convexity of the first portion of the urethra, before it is surrounded by the spongy substance." (*Ibid.*, n. 581.)

72. BOERHAAVE. "The urethra consists of two membranes with a cavernous substance between them. This substance is thickest between the end of the prostate and the union of the corpora cavernosa: afterwards it is thinner through the rest of its course; but thickens again over the end of the penis, is inflected outwards, and leaves in the middle the open mouth of the urethra in the glans. There it constitutes the exterior spongy surface of the glans, which terminates in a raised border or rim about the connexion of the prepuce." (*Inst. Med.*, n. 654.)

73. MORGAGNI. He mentions having seen numerous minute ducts and foramina in the urethra. (*Advers. Anat.*, I., n. 10, and seq.; IV., anim. 16—20; &c.) And he observes that the primary seat of gonorrhæa is in the larger ducts or canals, and that the pain felt is not unfrequently referable in the first instance to the spot where the greatest of these orifices is situated, about two fingers' breadths from the orifice of the urethra: this he found from the experience of several of his patients during the early stages of gonorrhæa. (*Ibid.*, IV., anim. 9.) He also speaks of other little canals which run into the larger ones (Index, p. xxx.), and of the glands described by Terraneus (*Op. cit.* IV., anim. 7).

74. As *the prostate is prefixed in a general manner to the urethra at its beginnings, so it is succeeded by that thick substance termed the bulb, stretched out in front of the spongy or cavernous substance of the urethra, with which the prostate, and the prior or membranous part of the urethra communicates, and by the bulb is connected continuously with all the substance down to the glans: likewise by the same with all the substance of the penis or corpora cavernosa; which shews that the bulb is the means of union and communication between all the parts that there are in the whole of this member. It does not at once encircle the urethra, but pyriform in shape, runs from the region of the prostate to its seat, where it unfolds, and surrounds the beginning of the urethra. Its motive or muscular membrane it takes from the accelerator muscles, whose fibres it afterwards transmits, as well outwards, as forwards and backwards, and on every side, to the corpora cavernosa.* In this place the whole active force of the prostate and the neighboring parts, and of the entire urethra, sheds itself over the substance of the whole member; more immediately over the urethra itself, to which that muscular fibre is adherent, and almost enters into the membrane, stretching round about it. In order then that no part of the bulb may be out of the reach of these invading forces, this muscle is attached to the septum of the bulb, which divides it into two, and from which it enters into the whole of its substance, and from this substance into the whole penis and its spongy substance; for it is attached to the suspensory ligament, where all the forces are concentrated; and at the same time in a particular manner it pours forth into that substance. *Wherefore the bulb is the terminus or goal whence issue all the active forces, both those from the prostate*

and vesiculæ, and those from the cavernous canal of the urethra and from the glans. Afterwards from this goal there is an extension or opening into the whole circumjacent substance, which determines and concentrates all its vibrations towards that common band under the urethra, as it were towards a genuine axis. Nature is always like herself. Where she pours forth forces, she determines them always to a central axis. So it is in the whole body, in the several muscles, in the testicles, and everywhere else: see above. Not otherwise does she distinctly institute any of her games. She passes from particulars to generals, from generals to most generals; thus from one degree to another; in other words, she passes according to order. The determinations of forces are to be inferred from the continuity of fibres and substances; for there are no forces where there are not substances. Thus we have to judge from the connexion of substances, of the determinations of forces by the urethra and around the urethra. *Thus the bulb is a kind of bridge that unites the testicles, vasa deferentia, vesiculæ seminales, prostate gland, caruncula, and membranous part of the urethra, to the several parts that succeed them and lie around them. And this, by means of fibres and membranes of different orders. All the modificatory vibrations then, both those that pass from the antecedents, and those that pass from the consequents, continually pass by this bridge distinctly into all the neighborhood; and by its means therefore, no part of a vibration or modification once begun can fail to arrive at consciousness.* The same vibrations indeed pass from all points of the urethra, of its membrane, its fibre, and from the vessels round about, into the substances of the member or penis, but in a general manner, and indistinctly: to produce distinctness, and therefore conformity to rules, and stability or enduringness, a starting-place, a field or course, and a goal, are requisite. The starting-point is the bulb; the field is the whole member or entire penis; the goal is the ligament under the urethra, where the three corpora cavernosa combine: not to say anything of several secondary parts, and particular starting places, goals and fields, for the exploration of which ages would barely be sufficient. These however are not difficult to recognize by a general law. For what nature is in great things, that is she also in lesser and least, with a difference of perfection in form, determination, and every quality and effect. *It is then the more necessary that all things should be done in this place with the utmost distinctness, inasmuch as upon this depends an effect so necessary as the generation of the species. And this effect requires the suitable disposition of all the parts, which here is only obtained by a subtle and imperceptible vibration of all, and the opening, constriction, and extension consequent thereon.* But of these subjects we shall treat presently with greater clearness when we come to speak of the corpora cavernosa.

75. *Humor and mucus in large quantities flow into this seminal and urinatory canal, not only from the prostate gland, but also from several other glands of considerable magnitude between the two membranes or cavernous substance of the urethra; and this humor and mucus unite especially with the seed as it escapes. For vast numbers of excretory ducts and lacunæ are visible, and innumerable lesser orifices, which are scarcely to be seen by the microscope unless the urethra be stretched.* Cowper found three glands of the size of cherry stones, and each with an excretory duct; and Morgagni confirms his account *(Advers. Anat.,* IV., anim. ?), although others have looked in vain for these glands: Littre moreover found another gland almost environing the circumference of the urethra. Terraneus and Morgagni describe innumerable and nearly imperceptible orifices through which a continual moisture transudes: lacunæ too are seen of divers forms, elongated, oval, rounded. *For where there are not glands present grouped into a visible mass, still every artery, when it has filled its vein with blood, and the blood cannot get out through any other artery, remains in a state resembling inflammation and obstruction: and in this state, agreeably to the universal custom in the other members and muscles, a copious mucous serum is excreted; which is determined through the minutest ducts: either into a larger common duct, or immediately into the grand duct of the urethra. Thus if there be indeed no glands, there are nevertheless as many corresponding springs of mucous serum as there are extreme serous ducts proceeding from the arteries.* These numerous orifices or oscula can originate only from the little arteries and their minute spires and whirls in the fine cavernous substance; for the same thing occurs here as in obstructed and inflamed members and muscles. The fact that an obstruction

does exist, and of such a kind that the arterial blood cannot pass elsewhere than into the veins, and that the venous blood cannot get discharged at all, is a point which we shall shew presently. Glands are not necessary: they may be present, or they may not; inasmuch as these springs or sources may be multiplied, and abundantly compensate for the absence of glands. *A humor of this kind appears to be much more calculated for the use of the outpouring seed than of the urine; and for lubricating the parietes, and protecting them from the acrimony of the urine. For it has scarcely any power of coming forth when the penis is relaxed and flaccid: on the other hand, when the urethra is on the stretch, the cavernous substances erected, the vessels and excretory ducts standing out and forward, and the orifices and the lacunæ drawn wide open, then it rushes out, and sometimes visibly runs before the genuine seed.* If glands be present, they are still only called into action during the general erection of the member, or only when the neighboring vessels and fibres, including the muscular fibres, begin to act, which does not happen when the urethra is relaxed. In this latter case, the ducts are folded together, and block up their own passages: moreover the lips of the orifices or lacunæ fall together, and produce a kind of semilunar valve by the meeting of their borders. Facts concur to prove the same thing; for when the penis has been long on the stretch, and the ardor of desire is felt, a moisture oozes out, which in like manner with the humor of the prostate runs before the seed. *Throughout the whole length of the urethra, this humor besprinkles the seminal humor, and not only produces the sufficiency of quantity therein, and serves it as a vehicle, but also diligently hinders the seminal humor from being poured around and leaping up against the walls of the tube, and being absorbed by the very copious veins and little oscula all seated therein; obliging it at the same time throughout the urethra to maintain the middle axis. For whatever is purest and most noble is also innermost, and occupies the centre: in this the case is the same in the general as in all the particulars.* This is a common rule in universal nature. Order itself prescribes it, as may be seen in our Doctrine of Order. If the cause is constantly to give forth the effect, then of necessity this noblest essence must keep strictly to the central axis, as well when it passes through the urethra, as when it glides into the uterus. Otherwise, were it thrown forth to the circumferences, it might easily be lost, but would be very uncertainly efficient. Besides which, the purest semen, of its own accord, does not run out to the circumferences, for it beats all the other essences in mobility: and hence to prevent it from being determined elsewhere by necessity, it is sprinkled over with humor from the very first orifice of the urethra, from the caruncula; which humor forbids its approach to the parietes, as we shewed above when treating of the humor of the prostate. The same fact is still more visible in the several arteries or arterial canals: the better blood keeps close to the axis; the viler blood and the serum beset the circumferential parts of the tube. See Part I. *Moreover this mucous humor attenuates the seminal humor in case the latter be too thick, and inspissates it if it be too thin, imparting to it the proper consistence and temperature, to enable it to serve its purpose in the womb.* This humor is not watery, but slightly viscid and coherent, or of a middle nature between glue and pure water; and so it tempers the too thick seed, which has been long kept in the vesiculæ seminales, and has got too dry; and also condenses the more fluid seed, which has not yet been ripened by remaining in the vesicles sufficiently long. It is clear that this humor approximates in its character to that of the seminal humor, for venereal gonorrhæa poisons this juice in the first place, afterwards attacks its neighbor, the blood; and lastly the seed of the vesiculæ (see Morgagni), which would not be the case were there not an affinity between them.

76. *There are innumerable impulsive causes that urge on the progress of the seed and humor to the outlet; for there is not a fibre, a bloodvessel, a part of the membrane, an excretory duct leading into the urethra, nor a part of the humor or of the seed itself, that does not increase and determine the motion to an ultimate issue in ejaculation or emission.* The facts shew that it is no simple transmission, but ejaculation, which takes place here: the manner however in which the several parts contribute to produce this effect, will require to be demonstrated of the parts one by one. *The whole venereal affection or desire induces a certain universal change both on the nervous and sanguineous systems, consequently on the body at large, and particularly on*

the organs devoted to generation. For it not only opens the several fibres, but also brings them into a kind of most subtle and pure vibration and modification, imperceptible however by reason of the exceeding swiftness of its moments. This modification is of an intimate order, and invades the simpler fibres, consequently all those fibres of which not only the testicles are composed, but the vesiculæ seminales also; as well as the membranes of the urethra and the cavernous substances. To this is superadded another and grosser modificatory vibration, which results from the touch undergone by the gland and its villous cuticle and papillæ, and which insinuates itself into the urethra and likewise into the corpora cavernosa. Since then all the substance of the penis and urethra, and of its appendages, is carried away into a similar vibration, so also the genuine sperm, and the seminal humor formed of it, is carried away into a similar vibration or intestine motion. This is the reason of the state of activity of the whole seminal material: for according to the state that lies in the tunics and their fibres, is the state that is communicated to the fluid they enclose; for state and state, continent and content, are exact mutual correspondents. Of these subjects we shall treat more fully in the sequel. *When this motion is put into the fluid parts by the causes aforesaid, a determination is required, or a force determining these parts to their outlets. For as the fluid parts are naturally constituted in intestine motion, nothing more is requisite than a determining force, to enable them to rush on with the impetus they have already got. Now the determinant forces are as many as the excretory ducts, all which run from the prostates in the direction of the length of the urethra towards the slit or orifice at its termination. The two seminal ducts direct their course thither; the ducts of the prostate pass around the caruncula into the cavity of the urethra: the ducts of the other glands stretch forth longitudinally through the membrane of the urethra for some inches, and where they issue forth, they pass obliquely, and end for the most part in oblong lacunæ and canals. So also the other innumerable and less perceptible ducts, which as they unanimously sprinkle their humor into the urethra, so they determine it onwards or forwards. Thus the forces determinant in the direction of the urethra are as many in number as the orifices or oscula, and as the drops that descend from the oscula.* Anatomists have shewn, and dissection proves it, that these ducts run for the distance of an inch under the membrane of the urethra, before they form their apertures. The ducts of Cowper's glands are two inches long. The other smaller ducts also run in a similar direction, as Morgagni was enabled to shew clearly by inserting a bristle into them. The lacunæ likewise attest this by their oblong form. Hence all the humor expressed from them is carried very obliquely into the canal, and so every particle of it determines to its direction forwards. We may conceive thus much of any mucous fluid running through a canal, —that if there be innumerable points that urge the volume to one quarter, it cannot fail to be determined thither: the force moreover is greatest if it runs in not so obliquely, but almost horizontally. From the forces thus multiplied over the whole circumference flows with certainty the determination of the volume, which is so constituted in the effort of flowing that nothing is wanting but the determination. *All this humor that flows into the urethra, rushes in with a certain force, for it does not burst out of its ducts until the reactive force is conquered by the active force. This was demonstrated of the humor that issues from the vesiculæ seminales and the prostate; it is the muscular coat and the reciprocal act that at length overcome the resisting force. The same thing also occurs in the several ducts and tubules that creep along the membrane of the urethra from the glands or arteries. While the urethra is extended, and those ducts drawn out, they are not opened until a superior force is provided, which is strong enough to open those little canals. Thus wheresoever any humor escapes through the urethra, it is pressed out, and by a kind of overwhelming force urges the surface of the little volume; and thus from all points thrusts and impels it to flow out not slowly but quickly, that is to say, causes its ejaculation.* The less power is required for this end, inasmuch as the sperm itself, intimately contained in the seed, and keeping the central axis, is volatile and moveable on its own account. *This determinative and impulsive force is increased according to the magnitude of the volume, that is to say, near the extreme fissure of the urethra, which likewise resists a little; for it appears to be surrounded by muscular fibres. Wherefore with all these causes to urge it, the semen may be said to be not simply transmitted but ejaculated.* The cavity of the urethra is larger and more capacious at the beginning; afterwards it decreases in size: towards the gland however it again enlarges, and afterwards contracts about the outlet: evidently in order that it may burst through this door with the more violence, and this, in the ratio of the magnitude of the volume that is pressing forth. There is no muscular compression of the urethra, but rather a yawning, together with extension: which shews that this forcible eruption is produced by more hidden causes; namely, by those that produce the forces of speed and fluxion in similar bodies and fluids. Respecting the active force which the urethra possesses, derived from vibration, we shall have to confirm its existence presently. All this force coming from the entire bulk and thickness of the penis is concentrated upon the urethra.

CHAPTER VIII.

THE PENIS AND CORPORA CAVERNOSA.

77. Heister. "The cuticle and cutis are the common integuments of the penis. The prepuce is a duplicature of the skin, which covers the glans. At the lower part of the prepuce is the frænum. Both are beset with nervous papillæ, and with glands, or more properly speaking, follicles, . . . which have been termed glandulæ odoriferæ. The coat or tunic of the penis is proper to it, and is robust and tendinous, and surrounds the other substances of the penis. It is sometimes double, having a cellular layer between its two parts, and which is conspicuous after inflation and drying. The two corpora cavernosa or spongy bodies constitute the chief bulk of the penis. They arise distinctly on each side from the ossa pubis, from a kind of peculiar beds; afterwards joining, they extend all the way to the gland. If liquid or air be driven into them, or mercury be injected, the penis is rendered rigid, and swells up. There is a septum or partition between the two corpora cavernosa; it arises from the line of junction between the walls of those bodies. Behind, it is comparatively thick, thinner in front: and it is perforated like a sieve. The penis is joined to the synchondrosis of the ossa pubis by the ligamentum Vesalii. *(Comp. Anat.*, n. 229.) In one subject, besides the suspensory ligament, I found two lateral ligaments of the penis, very similar to the suspensory: they arose from the ossa pubis on each side close above the origin of the erector muscles, and were inserted into the sides of the corpora cavernosa penis above the insertion of the same muscles. But I have found these muscles only

twice since, and much smaller than in the first instance, and hence I am unable to regard them as ordinary structures. *(Ibid.,* not. 28.) The muscles of the penis are various; they are principally of use for its erection. . . . Its vessels are wonderful in their quantity and distribution. . . . Its glans, head or balanus . . . has a smooth, polished surface, and is very sensitive by reason of its nervous papillæ. . . . Beneath the termination of the urethra . . . is the frænum of the penis. The posterior circumference of the gland is the crown or corona, underneath which lies the neck. Some anatomists state that there are little glands on these parts, . . . but Santorinus says that there are little orifices or ducts. The substance of the gland consists of epidermis and corpus cavernosum, which is continuous with the urethra." *(Ibid.,* n. 229.)

78. Winslow. "The integuments which cover all these parts are three or four in number. The first is the skin with the epidermis; the second is the ordinary cellular membrane, which in this place seldom contains any fat; the third is termed nervous; and the fourth is a particular cellular membrane, which is not always to be found. The first of these integuments, the cutis or skin, is a continuation of that of the pubes and scrotum, and it adheres to the second all the way to the basis of the gland, where the second ends. The rest of the cutaneous integument covers the glans without adhesion, and terminates by an opening. This portion is the prepuce. Along the whole lower or back side, both of the whole integument in general, and of the prepuce in particular, there runs a fine suture, which is a continuation of the raphe of the scrotum and of the perinæum. The inner surface of the prepuce is lined with a very fine membrane from the opening all the way behind the base of the glans. The same membrane is folded from behind forward round the glans, forming the proper integument thereof, and covering very closely in the way of an epidermis its villous surface, as far as the orifice of the urethra, where it joins the membrane that lines the inside of that canal. This proper membrane of the glans, and internal membrane of the prepuce, form conjointly along the flat part of the glans, from its base to the orifice of the urethra, a membranous duplicature, which like a septum or mediastinum divides this part into two lateral por-

tions, and limits the motions of the prepuce; for which reason it is called the frænum or bridle of the prepuce. The surface of the internal membrane of the prepuce discharges a fluid which prevents it from adhering to the glans; and perhaps serves likewise to dilute that which is collected at the base of the glans, and which comes from the glandulæ sebaceæ. . . . The third common integument, improperly called tunica nervosa, is of a firm, elastic, ligamentary substance, and its fibres are sometimes of a yellowish colour. It invests the corpora cavernosa and urethra from the glans to the symphysis pubis; and at some distance from these bones, it forms on the upper groove of the corpora cavernosa, a close duplicature; and by this duplicature, a flat broad ligament, whose plane runs directly upward, and is inserted along the beforementioned symphysis, as far as the tendinous base of the pyramidal muscles of the abdomen. This ligament has been called ligamentum elasticum, because it yields and recovers itself; and suspensorium, because it suspends these parts by means of its insertion in the symphysis. It sends off a detachment or ala on each side, one edge of which is fixed between the triceps muscle and the corpus cavernosum, and forms the ligamentary expansion in which the dartos is inserted. . . . It seems likewise to send down another elongation directly to the perinæum as far as the anus. The fourth integument of these parts is the tunica cellulosa of Ruysch, which immediately surrounds the corpora cavernosa and urethra, lying between these three columns and the third integument, from which it seems to be distinguished only by the closeness and fineness of its texture. It is sometimes hardly perceptible. . . . The musculi ischio-cavernosi lie along the roots of the corpora cavernosa; each of them being fixed by one extremity very obliquely to the inner lip of the ramus of the ischium, from the tuberosity upward. Thence it accompanies the root of the corpus cavernosum all the way to the symphysis pubis, and is fixed by its other extremity in the corpora cavernosa near their union: where the fibres of both muscles meet, and are reciprocally expanded over both corpora. They lie a little lower and more interiorly than the roots of these cavernous bodies. I have shewn two other and accessory muscles . . . fixed lower and more interiorly in the os ischium than the former, which they accompany all the way to the corpora cavernosa, and then leaving them, they are inserted chiefly in the urethra, near the bifurcation of the musculus bulbo-cavernosus. (*Exp. Anat., Tr. du Bas Vent.*, n. 558—570.) The corpora cavernosa are two very supple ligamentary tubes, laterally united to each other through a great part of their length, and closed at their extremities, two of which are connected together, and rounded like the end of the finger; the other two divaricate, like the branches of the letter Y, and diminishing gradually in size after the divarication, terminate in an oblique point. These divaricated and pointed extremities may be called the roots, and the round ends the heads. These bodies are almost cylindrical, . . . from the roots to the heads. . . . The ligamentary substance of their sides is elastic, composed of fine close fibres, which are partly transverse, and partly more or less oblique. The cavity of these ligamentary tubes is entirely occupied by a very strong cellular or cavernous tissue, which seems to be only a continuation of the substance of the sides. These cells communicate with each other, and are always more or less full of blood, resembling pretty much the cellular substance of the spleen, but with this difference, that the sides of the cells are thicker in these cavernous bodies, and their cavities have no accessory tissue. By the union of the two corpora cavernosa externally, two grooves are formed, one above, the other below. The lower groove is somewhat broader than the upper, and it is occupied in its entire length by a third tube, . . . namely, the urethra. . . . The roots of the corpora cavernosa are fixed respectively to the edge of the small ramus of the os ischium and os pubis. They meet towards the symphysis pubis, where each of them becomes a cylindrical tube, and unites with the other in the manner already said. The heads or rounded extremities immediately join the base of a distinct body, the glans, which is an expansion of the urethra, and closely united to it. . . . The apposition of the corpora cavernosa from their roots to their round extremities or heads, forms a particular septum through the union of the transverse fibres of each. Between the fibres of this septum, several small void spaces are left, by which the corpora cavernosa communicate with each other, and therefore by blowing into one of them, we presently inflate the other. Toward the rounded extremities, the septum diminishes every way. (*Ibid.*, n. 523—529.) The arteries of these parts come principally from the internal iliac or hypogastric arteries, and some also from the external iliac and the crural. The principal arteries are the internal and external pudic. The external pudic sends a branch to each side, which having passed out of the pelvis by the side of the sacrum, runs on the inside of the tuberosity of the ischium to the roots of the corpora cavernosa, along the inside of the musculi ischio-cavernosi or erectories. It sends ramifications to the bulbous head of the urethra, and to the corpora cavernosa; and with its neighbor the glutæal, with which it communicates on its passage, it likewise supplies the scrotum. The internal pudic having supplied the rectum, bladder, vesiculæ seminales and prostates, and communicated with the hæmorrhoidal vessels, pass under the arch of the ossa pubis, and partly enter at once the corpora cavernosa, partly run along their upper side, sending off small lateral branches, which surround these bodies, like irregular half-arches, and penetrate them by numerous ramifications. The crural arteries send each likewise a considerable branch which . . . is distributed to the integuments of the penis, under the name of the external pudic, and which communicates by lateral ramifications with those of the internal pudic. Communications exist not only between the internal and external pudic of the same side, but also between those of both sides, which anastomose with each other. The distribution of the veins follows nearly that of the arteries. . . . The principal vein is that which passes directly under the symphysis pubis between the two arteries, and runs along the whole upper groove formed by the union of the corpora cavernosa. It is very large, often double, seldom triple; but the trunks do not separate while in the groove. This vein has a number of valves. The great middle vein is formed by the union of the hypogastric branches, which after passing on the two inner sides of the pelvis, meet about the middle of the arch of the ossa pubis. At this place we observe a venous plexus, which covers the upper convexity of the first portion of the urethra, before it is surrounded by the spongy substance. . . . The nerves of these organs come from the lumbar and sacral nerves, and communicate with the great sympathetic and the mesenteric plexuses. Near the arch of the os pubis on each side they form together a particular cord, which passes under that arch along the upper side of the neighboring corpus cavernosum, near the artery already mentioned. In their passage over the corpora cavernosa, they send off a great many branches, which surround these bodies on all sides. They run between the skin and ligamentary integument; being so disposed, that the arteries lie between them, and the great vein in the middle. They must be examined presently after the skin has been raised, because when the ramifications are dried by the air they disappear. . . . There is likewise one nerve on each side, produced from the union of the second, third, and fourth pairs of the sacral nerves, but especially from the third; it goes out of the pelvis above the ischio-sacral ligament, passes by the inside of the tuberosity and small branch of the ischium, and is distributed to the corpora cavernosa, to the muscles belonging to them, and to the neighboring parts." *(Ibid.,* n. 576—588.)

79. BOERHAAVE. "The whole body of the penis formed of the corpora cavernosa is surrounded by a cellular membrane of a very remarkable structure, which also goes round the middle septum: in the next place it is invested with a firm nervous envelope which sets bounds to its expansion; and last of all, by the cutis and the cuticle. The arteries . . . carry the blood into the corpora cavernosa of the urethra and penis, where dividing into innumerable capillary arteries they are continued into the little veins. In the latter there are innumerable hollow cells which communicate with each other, and all discharge themselves into one great vein, which runs down as a single trunk along the back of the penis under the ligament connecting the ossa pubis; and which vein near the prostates divides at length into two branches that empty themselves on both sides into the internal iliac veins. The veins however of the corpus cavernosum of the urethra are placed beside the accelerator muscles to serve their end. . . . The muscles arising from the external tuberosity of the ischium below the origin of the corpora cavernosa, are inserted into the firm nervous envelope of these bodies, and there disappear in tendons. . . . The commencement of each of the corpora spongiosa must be more closely pressed up against the tuberosities of the ischia by the

muscles in this situation swelling, and their veins must be closed. The body of the penis also must be pressed more than at other times against the ossa pubis and their middle ligament. The great vein of the penis must likewise be compressed, but the veins of the prepuce less. The arteries therefore will be filled as well as the veins, and hence the veins will be unusually compressed. Hence again the cellular sinuses will be filled with blood." *(Inst. Med.*, n. 655—657.)

80. MORGAGNI. "When the corpora cavernosa of the penis are inflated, that of the urethra is not distended. *(Advers. Anat.*, Index rerum, p. xxiv.) The septum, as it is called, is not a distinct part from the corpora cavernosa of the penis." *(Ibid.)* On injecting the artery which enters at the side into the bulb, our author found that no part whatever of the spongy substance of the urethra was distended by the injection, and that it found no passage into the nervous bodies of the penis; but the water . . . came back through the vein that wound about among the integuments of the penis. The substance of the nervous bodies of the penis appeared to be tinged with the green color of the injection. (See *Ibid.*, Advers. IV., anim. V.) "The suspensory ligament of the penis . . . arising in the first place sometimes from the linea alba of the abdomen, much more often from the middle of the os pubis, or still higher up, at other times a little lower down, and closely connected in front to the cartilage by which those bones are united together, stretches forth in the manner of a firm and powerful membrane to the subjacent dorsum of the penis. . . . It is by no means . . . fixed into the septum of the penis, but unfolds into a strong tunic, which invests nearly the entire penis, and moreover the scrotum itself underneath the skin. Expanding posteriorly, it also covers those muscles which are sometimes called the dilators of the urethra; while in front it forms what is termed the panniculus carnosus of the penis, or if the reader prefers to say so, is continuous with that structure. At length passing to the scrotum, it constitutes both the dartos and the . . . septum scroti." *(Ibid.*, I., n. 18.)

81. SCHURIG. "The various names of the penis, as virga, colis, veretrum, &c. *(Spermatologia*, cap. III., § 1, p. 89.) Cases of the penis being stiff in dead subjects, one from Fred. Ruysch;

for the penis is erected by flatus or air blown into it. Ruysch, by means of a small tube introduced between the cutis and the cellular membrane which is commonly termed the dartos, inflates the scrotum. His erection also of the male organ itself. Erection has frequently been observed in the bodies of persons killed in battle in the summer season, and left lying on their backs. *(Ibid.,* § 10, p. 101*s.*) Cases of satyriasis and priapism. *(Ibid.,* § II., p. 103.) Monstrous members; little ones; shrinking and loss of the penis by poison, by congenital defects, by gangrene, cancer, excision; cases of entire want of penis. *(Ibid,* §§ 16—24, p. 109—123*ss.*) Cases of crooked, double, bony members: of the penis growing to the abdomen; of monstrosity of the glans. *(Ibid.,* §§ 30—38, p. 128—135.) Wind escaping through the penis, in Schenck., Observ. Med., lib. 4 fol. m. 526. Many examples from Thomas Bartholin. *(Ibid.,* § 41, p. 138.) Imperforate glans." *(Ibid.,* § 46, p. 142.)

82. Ruysch. "If the nerveo-spongy substance be much and strongly inflated, the veins that meander through the dorsum of the penis, and whose twigs supply the nerveo-spongy bodies, are inflated at the same time. *(Obs. Anat.,* Amstel., 1721. C. p. 96.) The author remarks that the erection of the glans depends upon the proper constitution of the arterial vessels. All the veins distributed through the penis, or if not all, at any rate all that I have met with, are perforated with large pores and visible foramina, like a sieve, in the same manner as we see in the splenic vein of the calf; though this has not been hitherto observed. This perforation of the veins is the reason why the blood returning through the penis can so very quickly go back from the penis, which latter gets flaccid in a moment. The spleen bears a considerable resemblance to the penis. The internal veins of the penis are extremely difficult to exhibit: many anatomists deny their existence. These veins however were inflated by the author (Ruysch), though the substance of the inner vein, from its porous character, may be very easily confounded with the intrinsic and highly porous substance of the penis." *(Ibid.,* p. 97*ss.*)

83. *Every affection of the sensory of the cerebrum, every passion, as it is termed, induces a peculiar and answerable state on the fibres of the whole body. This state is first induced on the internal sensories; and since the medullary fibres of the cerebrum, and the nervous fibres of the body, arise from these sensories, hence a similar state is induced on the fibres; consequently on the whole body, which is woven organically of nothing but mere fibres.* On these subjects let the reader consult Part I., and we would also refer him to Part V., in which we shall treat of Joy, Sorrow, Anger, Mercy, Love of different genera and species, &c., &c. *Thus Joy expands, dilates and gladdens the fibres, but sorrow and grief constrict or shrivel them. Anger hardens and mercy softens them. Love, and sexual love especially, is a kind of concentration and exaltation of all the delightful affections—of joy, mercy, and many others like them; and it not only expands but extends the fibres, and brings them into a state of the utmost modificatory tremble,—a state in which there lies an effort to pour out all the essence, spirit and life whatever that they contain, and to transfuse it into a second self; and so by this means to transplant all its own into another being, to such an extent that its own possessions seem its own no longer. This is manifested by the exceeding violence of the love towards the beloved object; by the storgē or parental love towards little children; by the small value set upon life whilst this love occupies the sensories: likewise by the nature of sexual desire; by its intimate delights before the act, and its supreme pleasure in the act.* These are points belonging to Psychology. *As is the state of the fibre, such also is the state of the purest spirits in the fibres, and such is the state of the spirits in the blood, nay wherever they reside.* This is clear enough from the signs of love in the very eyes, exhibited

in their visible radiation, and from the intimate life of the several parts, as well as of the blood itself. *This state being superinduced upon the fibres, a similar state is induced on the system universally; but it gives forth no effect or result excepting in the ultimate sphere of the fibres, that is to say, in the organs that are formed with an express view of giving forth this effect. In the latter then is planted a faculty for receiving those modes, applying them to the genitals, advancing them onwards, and ultimately to the effect, which is the expulsion of the seed.*

84. *When therefore all the fibres are put in this state, and particularly those last or ultimate fibres that construct the several organs of generation, then all the simple fibres that are present, and all the other fibres, compound, motive and tendinous; also the bloodvessels, and the membranes which they compose; the septa, the integuments, be they pure, or be they gross,—all put on a similar state: in a word, they all naturally expand and extend; that is to say, those which are lying flaccid become erect and stretch themselves forth.* This is the natural effect; for all the structures composed of fibres, put on the state of their fibres: and be it observed that there is nothing whatever in the universal body but is an organic structure or product of fibres: on which subject see the Part on the Fibre. *Therefore the cells of the corpora cavernosa (both that of the penis and that of the urethra and the glans), put off their state of relaxation, and become erect and tense; for according to the law, they are both expanded and extended. So likewise the vessels, and especially the arteries, which on other grounds are of such a nature that they are extended when they are expanded.* See Part I. *Thus when the minutest arterial vessels in the corpora cavernosa are expanded and extended, that is to say, erected; and when at the same time the like occurs to the larger vessels or trunks composed of similar fibres, then the blood flows in abundantly, strongly and pronely, and fills up the minutest particular arteries, which finish the work; and so manage it, as that when the several membranous septa of the cells are raised by their own force, they shall at length and at last be called forth or excited by the arterial bloodvessels. For this reason it is that we attribute the erection of the penis to the blood, when yet this is only the last effect, and although apart from the prior cause it will indeed erect the mem-*

ber, yet it does not always incite to coition from an internal cause. Ruysch has delineated with great accuracy the mode in which the large artery traverses and ramifies through the corpora cavernosa. It runs through the middle of the corpora cavernosa penis, and divides upon and ramifies over the several membranes of the cells: so much so that there is not a part of the membrane but is irrigated by those vessels. Furthermore an abundance of nervous fibre flows into them; and it was on this account that the ancients, and Ruysch likewise, termed these corpora cavernosa, nervous and spongy bodies. It follows from this, that the animal spirit principally is operative here: not by any means that it runs into the cells, but it flows into all those membranous walls that make up this spongy substance both in particular and in general. For the only effect is, that all the cellular parietes are erected, by the aid of the spirits and the blood, that is to say, by that of the fibres and the arterial vessels. *When therefore the penis is expanded and extended by the fibres and little arteries in particular and in general, all the avenues into the veins, and the veins themselves, are tightly closed, to the intent that the arterial blood alone in the meantime may be kept shut in its little beds. These veins are quite constricted and barred by the internal force, and particularly by the external, occupying the surface, as they do, like great hives; and when the corpora cavernosa are extended, they are narrowed both breadthwise and lengthwise to such an extent that the blood cannot escape into them.* The arteries are transmitted through the middle of the penis, while the veins are rejected towards the surface. The direction of the great artery and its branches is given in Ruysch's figure, and the reader will find in Heister *(Comp. Anat.*, tab. v., fig. 22, 23) a delineation of the veins. It appears from the plates, that the veins which occupy the surface cannot fail to be compressed on every side in the swelling of the member. Injection proves the same thing; thus when the arteries are filled, and the member tumefies, not a drop of arterial blood escapes into the veins. Reason too dictates as much; for were there a free circulation, the arterial blood would not tarry on the spot, and could yield therefore no result. And it contravenes experience to say, that both the arteries and the veins at one and the same time, and also the cells and follicles, could be filled and distended with blood. Moreover there are several lesser veins in the cellular structure of the penis. Of the mode in which these are constricted anatomy does not yet inform us; yet we may reason from the great to the small, or from the general to the particulars, at any rate so far as to conclude to the existence of a similitude and parallelism between the cases. *The penis may also be expanded and extended, that is to say, erected artificially, or by injections; by forcing in water or any other liquid, spirit, heat [?], mercury, or air, into the veins and the cellular substance of the corpora cavernosa, and for a brief space preventing it from running out again. This cause of erection is purely artificial, and not natural; for anything that expands the cells and their parietes both lengthwise and breadthwise, will cause the member to tumefy.* All the membranes, both particular and general, belonging to the corpora cavernosa, are erected naturally by the force of the spirits in the fibres and of the blood in the arterial threads: if these same cells are erected by the impletion of the cavities, the same effect will follow, though it will be only artificial. *It appears from the foregoing considerations, that the muscular fibre contributes not at all to that erection; for erection exists from desire alone quite antecedent to the action of any muscle whether belonging to the penis or to the urethra, and whether it be of the erectors or the accelerators, as they were called by the ancients.* For the penis is excited and stiffens by thought alone, by the mere sight of sexual objects, and the mere touch, before any muscle is in play. *But the muscular fibre, and the aponeurotic membrane both of it and the suspensory ligament; also the remaining sheet of this membrane, contributes thus much and no more—that it reacts against the action of the proper fibres and arteries of the penis and urethra; thus ensuring the due and corresponding reaction and action, and thereby the limitation of the expansion and extension: and moreover the further effect, of the active force superadding a stronger reactive force in the venereal paroxysm or act, so that not only the seminal humor may be pressed out from the vesicles, but also its companion from the prostates, glands and arterial clews in the urethra.* As indicated above. For were it not for the reactive muscular force, or the moving fibre, which is the grossest of all, the expansion, extension and action could not be limited. This motive fibre, consisting as it does of nothing but nervous fibres and arterial capillaries, is also at the same time extended; but still the muscle which governs it presides over the whole of that fibre, causing it to suffer itself to be extended and expanded more or less. This is the reason why its fibre extends over the whole penis, likewise also the suspensory ligament, to which at last the fibres are on all sides attached. They draw their force from the muscle. Tendinous fibre of the sort is not usually attached elsewhere than to the extreme parts, which it is wont to tighten and relax; thus by one extremity to the bones of the ischium and pubes; by the other around the whole penis. Were it not for such limitation the penis would swell beyond what it could afford, for there would be nothing to resist. The utmost force of action on the part of the muscles exists in the act, at which time the juices are to be squeezed out of their organs: now the excretory ducts resist them, and hence in order that reaction may then overcome action, a peculiar condition is required, by which the muscular fibre at this time dominates over the nervous and arterial fibres, and at that instant all is set free, or explosion occurs: but not naturally before this act, for then the active and passive forces are equal. But thus much in general: it will now be well to consider particulars: without a general idea however, the ideas of singulars would not easily be perceived, for they must attach their idea to some general. Such is the lot and condition of human intelligence.

85. *But it is not sufficient to shew the causes: it is necessary also to demonstrate the means that concur, and the mode in which they go on. Everywhere nature must proceed from the innermost to the outermost; if from the outermost to the inner and the innermost, the order is perverted. There are steps or degrees, and there is a ladder, by which nature descends and ascends. Throughout the descent* a posteriori *causes must concur with causes* a priori, *and both make up one series. This series it is which has now to be unfolded; otherwise we shall have no distinct perceptions.* On these subjects we refer the reader to our Doctrine of Order. Nature, in proceeding from the innermost to the outermost in substances, does the same in modes and other accidents: but matters are so arranged that there is

always some peculiar cause of the inferior order to call it into existence and play. *The innermost cause excites the simplest fibres. This is none other than the final cause, that is to say, in the present case it regards the procreation of offspring for the sake of society on earth and in heaven. This cause is put into the nature of the soul. Thus as the soul, which is in the inmost sphere, and acts spontaneously, regards nothing but end, and the effect itself as an end, so the first exciting cause of venereal love, is the procreation of one's kind. This, as lying in the purest eesence, lies there of its own accord, and sets in motion the very simplest fibres of all.* The reader will see marvels respecting the simple fibre in our Part on the Fibre. The simplest fibre is not that which passes visibly from the medullary part of the cerebrum into the nerves; but myriads of simple fibres make up one such medullary fibre: their mode of action will be shewn in Part V. This cause lies deeper or higher, that is to say, it issues from the nature of the very soul. *The second exciting cause is that which puts in motion the fibres made up of simple fibres,—which in short calls into play the animal spirit. This fibre or animal spirit is not called into action by the cause that excites the simple fibres, unless indeed in a remote way: for this cause does not flow into the compound fibre otherwise than as a certain faculty or a certain effort, into lower spheres. Hence, to call it forth, a proper cause is required, which afterwards breaks forth into act, as it were from a foregone effort. These exciting causes consist of thoughts and imaginations respecting love and amatory objects: the external causes consist of the sight of these very objects, of conversation, and still more of kisses, contact of various kinds, &c. These loves excite desire, which however does not deserve to be called rational, but only natural, being common to men with brute animals.* These proper causes excite the animal spirit and compound fibre to lechery or desire, whose effect is, as I before intimated, a modification of the spirits and of the compound fibres, which are brought together into a state of expansion and extension. *The third cause is the external contact, and particularly the embrace, and that of the glans penis itself, furnished as it is with such numerous sensitive papillæ, from the corona glandis to the head and the frænum where the papillæ are concentrated, and where the feeling itself in a manner converges. This exciting cause is the third in order, and stirs up the blood itself to concurrence with the prior causes.* Thus nature proceeds from the innermost to the outermost, or from the soul to the blood, which is in a manner the soul of the body; for the blood is the soul's vicegerent in the body or ultimate sphere. *The last exciting cause of venery, desire, or rather lust, consists in whatever excites the moving fibres or muscles; for the muscular fibre is composed of genuine fibre and of arteries; hence the cause that calls it into play is later or posterior still. This cause is excited by the succussion and agitation of the muscles of the penis, urethra, loins, abdomen, pelvis, or of the bones of the pubes and ischium. By this means the extreme degree of excitement is produced.* Everybody knows this. We shall shew presently that the muscular fibres are excited by this cause. *In order then that the first desire, or the first love, which is genuine, and exists solely with a view to the procreation of offspring, may come forth in act, it is necessary that all the causes do concur, and this, in regular succession from the first to the last: whereby a series of causes is produced, in which the innermost cause lies as the principle and soul of the rest.* As this is a subject pertaining to our reason, it is evident, nor do I know whether any one can doubt of this series. All these causes likewise concur, viz., to constitute an entire series, in the venereal act. The first, which is the origin of all, is not denied. Nor the second, viz., eyesight and its objects, also embraces, the whole of touch, and the various things that occur, and at once insinuate themselves into the imagination. Nor the third, in the act itself, when the glans, furnished with papillæ, feels the irritation of the objects that it meets. Nor the fourth cause, consisting in the actual agitation of the body and of the various muscles. If one cause in this series is wanting, the natural progress does not exist. *This series does not consist always of four degrees or terms. There is a series that consists only of three, and commences with sight and imagination, in which case we have concupiscence, which does not belong really to the soul, although it proceeds to its ultimatum or end in a similar career. Nay there is a series which consists of only two terms; or of the rubbing and provocation of the glans; but the prior link, or imagination must come in aid before the effect takes place; this how-*

ever is the inverse of the order of nature. This is because it begins from the posterior cause, and rises to the prior. It is thus far natural, that we begin *a posteriori* all that we do, and our order seems to be instituted in this wise, that we proceed *a posteriori ad priora,* before we proceed *a priori ad posteriora.* But as this is contrary to the order of nature, it is by no means conducive to the weal of life, but on the contrary hurtful. See above.

86. *But it is not sufficient to have recognized the causes: it is also necessary to know the influx of the causes into the genital organs. The first cause flows in a general manner into all the simple fibres of the whole body, and expands and extends them: also modifies the purest animal essence, keeping it in the mood of pouring itself entirely forth. This effect is only possible in the ultimate organs, viz., the testicles, which are formed expressly for this purpose. The above essence is not discharged anywhere else, unless perhaps into the compound fibres or into the animal spirit, and by this at last into the blood.* That the fibres of the testicles are of the simple order, or that the organism and structure of the testicles is such that the simple fibres end in them, and there emit the purest essence, may be seen in the Chapter on the Testicles. The fibres likewise terminate in the epididymides and vasa deferentia; but here is the last of this same cause; for every cause has its maximum and its minimum: and already in the epididymides and vasa deferentia this purest essence has grown up into the first essence of its seed. *The second degree, however, considered in itself, does not act immediately upon the testicles and epididymides, but upon the vesiculæ seminales, whose fibres it expands and extends; and also modifies them, enabling them to act correspondently on the fruitful seed itself or that of the first degree. Thus the effect comes to pass with the aid of the intestine motion of the primitive parts of the seed, and of the modificatory motion of the fibres of the second order.* But on these subjects see the Chapter on the Vesiculæ Seminales. *The third degree however, considered in itself, acts upon the prostates, and upon the bulb and all the glands of the urethra; also upon all the walls of the cells of the corpora cavernosa both of the penis and urethra; and expands them, and constringes them; and this, by a still more evident modification: it*

likewise acts on the fibrous and muscular coats of the arteries, enabling the blood to run into the now erected septa, and to complete the work thus far advanced. The facts of the case shew clearly enough, that this degree expands, extends, and modifies, not only the compound fibres, but their groupings and tunics. These septa are comparatively large, and the corpora cavernosa consist of nothing but septa of the kind. The finer then the cellular or spongy substance becomes, the swifter and surer is the effect. Thus when the cause proceeds *a priori ad posteriora,* the cavernous tissue of the surface is erected in the first place *a priori;* then afterwards the inner, which is more coarsely cellular. But if on the other hand the cause come *a posteriori,* then the inner cellular texture is erected before the outer. First of all, if we proceed in order, the erection takes place in the spongy substance of the urethra, which is denser and finer than the other sponges. For it proceeds through the nerves and through the arterial vessels; and the nerves run at the side of the arteries, and insinuate themselves into the several septa; they also follow the artery and its branches. *Nerves cannot be wanting to enter all the membranes, general as well as particular; nor can arteries be wanting to irrigate the several septa; for they traverse the middle of the corpora cavernosa, and on several hands a way is open for them.* The nerves come principally from the sacrum, being nerves of the cerebellum; likewise from the intercostal and par vagum, or from the mesenteric plexuses, of which the pelvic plexus is formed. The arteries are from the iliacs, and from the internal and external hypogastrics, from which at this time a free influx is allowed. *The fourth degree excites the muscular fibres, that is to say, those of the muscles belonging to the penis and urethra, as well as those of the abdominal and all the lumbar muscles; excites likewise the suspensory ligament, which is extended over the body of the penis together with muscular and almost tendinous fibres; and individually over all the organs preceding the penis, as the prostates, the bulb; and also distinctly into the spongy septa. Now the muscles thus excited compress these parts by alternate actions, compelling them forcibly to ejaculate their stuff, the reagent force being overcome.* With regard to the muscles, the erectors as they are called are proper to the penis, and they extend over the whole member in general:

the accelerators however extend over the bulb, and from thence on every side into all the cavernous substance of the urethra and the penis. That therefore which the erectors effect in general, the accelerators effect upon the septa of all the particulars.* These muscles which surround the prostate and the bulb, with extreme force urge and explode the whole of the contents: thus the motive fibres of the penis threw out all the juice that is discharged immediately into the urethra. The suspensory ligament likewise, which goes over the tunic, and constitutes its tendinous portion all the way to the corona glandis; and is continued almost from the linea alba of the abdomen, also over the dartos, and to the septum scroti. This tunic being tendinous must necessarily be reactive, but general; whilst the motive fibres of the muscles before-mentioned enter the septa particularly. Respecting the suspensory ligament, see Morgagni. *Moreover the great ischiadic or crural nerve is now in full work*—in maximo opere—*and goes to the muscles of the loins; and on the action of this nerve, consequently during action on its natural situation, depends the effect, not only of the work, but of the body also.* All the lumbar muscles depend upon the great ischiadic nerve, the largest of the nervous trunks: a particular nerve also going to the penis and vesiculæ seminales, traverses that nerve: see Winslow. This being the largest nerve, and being stirred correspondently with the whole of the natural motion of the body in the act of coition, it follows that we ought to be in the natural position when that work is being performed; otherwise, as when in the erect position, that nerve grievously labors, and is distressed, producing the effect of lassitude and debility upon the muscles of the loins and feet, which at this time are deprived of all their juice.

87. *There are two spongy or cavernous substances, one proper to the urethra, the other proper to the penis: these two are distinct, there being no mutual communication between them. The spongy substance of the urethra, and that which constitutes the surface of the gland, is the first excited by the causes of venery, and when excited, an influx of arterial blood takes place into the*

* The reader will constantly observe that the author imbeds philosophical terms in organization. This incarnation of philosophy has to be carried into every sphere; and then what was before abstract and useless will become naturally operant.—*Tr.*

spongy substance of the penis. It is found by injection and inflation, that the glandular substances of the urethra and penis are distinct, and that the substance of the urethra pours forth very largely—*crasse*—into the gland. The cavernous substance of the urethra is first erected in venery, for the próstate is prefixed to it, also the vesiculæ seminales and the bulb. *The following appears to be the order of erection. First in the venereal affection stands the spongy substance of the urethra and the glans; afterwards, but almost simultaneously, the superficial sponge of the penis, which is of a finer texture; lastly that which lies in the middle, and is coarser in kind. But a different and almost contrary order prevails, when the erection takes place, not from the prior cause of venery, but from some posterior cause; in this case the grosser part of the penis is excited first; next its finer surface; and at length and at last, the spongy substance of the urethra. In the erection of the penis apart from the penis*—in erectione penis absque pene—*as during the time of repose, in the night, during lying upon the back, lying hard and the like, nothing but the substance of the penis is erected, and a little only the substance of the urethra, which continues all the time in a kind of flaccid state; although from the expansion and extension of the penis it seems to put on the general state of the penis.* This follows clearly not only from an examination of the facts, but also from the connexion of the causes. The substance of the urethra immediately is erected by the state of the vesiculæ and prostate, and also of the bulb; consequently by the rush of the semen: it is also of so very fine a texture, that this state of erection of the cells is a consequence of this influx: but when erected or expanded, there is then an influx of the blood into the spongy substance of the penis, whose superficial substance, or that next the tendinous or aponeurotic coat, and which is considerably purer or finer, is erected. And as this is continuous with the other or middle substance, it prepares it for letting the blood rush in; for the blood does not flow into the parietes unless this substance be prepared by the extension of the fibres. But if love as a cause, or any other, extends the penis, it is a sign that the blood alone did it, and the relaxation of the fibres of the cerebellum by the fibres of the cerebrum. This again whoever chooses may experience in his own person. Thus the erection of the penis in sleep when no dreams or amatory imaginations are present is spurious and purely carnal: the other however is genuine: the former does not produce the latter until the latter is called forth by its own causes also: it is mere inflammation, as may indeed be known by the feel of the glans and urethra, which are flaccid, notwithstanding the stiffness of the corpora cavernosa penis. *There is also another kind of spurious erection, in which the veins alone and the cellular cavities are filled with blood:* this is sometimes observed after coition, and is recognized by the weight of the organ.

88. *As soon as the venereal frenzy has cooled down, the inmost organisms grow flaccid, and successively, the middle and the last or outmost. In this case the blood does not run back through the artery, but unloads itself abundantly into the inner and lesser veins, and afterwards into the larger veins and the venous channels; which are severally relaxed in the same successive order. At last it escapes from the arteries into that solitary vein near the os pubis; and so the blood passes out, and by degrees the penis collapses.* The stretching of the cells is what causes the communication between the arteries and veins to be suspended; hence, as soon as the cause of erection ceases, the communication is reopened. The veins moreover are provided with valves to prevent the blood from flowing back; but in proportion as the penis is unbent and relaxed, the great dorsal vein is opened, which is sometimes a double or triple channel. The blood which it carries is received into a larger or fuller channel outside the penis, and then splitting into two streams it flows back into the iliac veins. Such is the organization at the time, that it affords a liberal communication of the several streams, although previously that communication was everywhere precluded in the general as well as in all the particulars. *To prevent the blood of the spongy substance of the urethra and the glans from rushing out precipitately, and to cause it to escape with sufficient slowness; also to ensure the vein being exactly shut while the act is going on, it has to run down by wonderful plexuses about the os pubis and the upper membranous portion of the urethra: which shews clearly enough, that when that part of the urethra is extended, no portion of the blood can possibly get out.* The anastomoses of these veins are termed by Santorinus the

labyrinth, which is so placed and turned, that the blood cannot possibly run out during the act, for it is indeed superficial in this situation, and when the urethra is expanded it as it were disappears. *In order that nothing may resist the impetus of the blood as it is about returning, especially while a means of escape is still open outside the body of the penis or urethra, a passage is afforded also from the veins into the cavities of the cells, and from these again into the veins; for as in the spleen, so here, the walls of the veins are perforated with innumerable orifices.* As there is an influx and reflux, so the first outpouring from the veins appears to be into the cells, in order that the penis may be still kept a little expanded, and that the blood may not rush out precipitately, distending the vessels too much, and after a few acts of coition rendering them incapable of resisting. The penis is equally erected by the blood in the veins and cells, and by the blood in the arteries, as shewn by inflation and injections: see Heister's figure of the erect penis. Respecting the holes and perforations in the veins, see Ruysch and other authors: the cells also appear on dissection to be more or less loaded with blood: see Winslow. But this red blood is forced into the veins as the cells are relaxed, or as the parietes are approximated to each other. The fact that they can be closely approximated is plain enough during the relaxation of the penis; thus, in some persons, particularly in decrepit subjects, the penis is so far drawn back into the scrotum that it seems to have disappeared. Thus in order that but little room may be afforded during relaxation, the blood is expressed from the cells back again into the veins. The cells also communicate with each other, as likewise do the corpora cavernosa penis, and by this means it is provided that no drop of blood be anywhere detained, but that every such drop shall run into some veins through some perforation. Were it not for this communication between the cells, schirrosities and obstructions of the cells might very easily arise. *And in order that the cellular spaces of the corpora cavernosa may be perpetually irrigated by a certain humor, and thus be kept in a proper state for performing their function, an abundant liquid is drawn and expressed into the cells through the cellular tissue from the scrotum: and this for the further reason, that the venous blood which runs into them (the corpora cavernosa) may be diluted to prevent its coagulation or thickening. For the cellular tissue around the scrotum is continued on into the cellular membrane of the penis, and by the prepuce to the circle or corona: where it is reflected in the form of a fine cellular substance, and thence passes by continuity to all the cells.* Wherever there is any cavity that is expanded and constricted, there a suitable humor is present to water it. This is the case in the abdomen, the thorax, the pericardium, the cerebrum, &c.; the same rule holds also in the minutest parts, for the power of expansion is maintained by the humor. In order then that the cells under consideration may not be wanting in their proper humor, all that humor which surrounds the cellular texture of the scrotum (see above) is continued to the penis; yet it does not flow in directly, but by roundabout passages, that is to say, as soon as it has reached that duplicature: so that it would appear rather to be squeezed thither than to run in of its own account; for the tendinous or aponeurotic coat prevents it from passing elsewhere. But as to whether any portion of it runs into the spongy tissue of the urethra and the glans,—this is a question which we may guess in the affirmative; and in this case may suppose that the communication takes place about the frænum, where the prepuce is attached; although the cells of this substance are sufficiently irrigated by arteries of their own; whence the moisture excreted into the urethra. It remains still to be observed, whether or not it is this humor of the scrotum that also passes into the urethra, the scrotal humor thus collecting its streams around the urethra and around the semen; just as the humor of the testicles runs through the urethra. *If we are entitled to judge from the constant rule of nature, we may then conclude, that as the semen rushing out from the testicles keeps the central axis of the urethra, so the humor of the scrotum keeps the surface; with a view to the two being again concentrated, and meeting together on the ground of their principles. This connexion of things may also be confirmed by the external surface, inasmuch as a particular line runs down from the orifice of the urethra and from the frænum through the whole of the lower portion of the coat of the penis, and also of the scrotum, and at last traverses the perinæum to the anus: also inasmuch as there is a similar communication through the septum scroti. Moreover also the expansion of the penis com-*

presses the spongy substance of the urethra and glans in such a manner as to contribute to the expression of this humor. But all these subjects demand further examination. The connexion of things, apart from the confirmation of experience, leads us in this direction. The investigation, however, will not be difficult. The compression of the spongy substance of the urethra by the corpora cavernosa by virtue of their situation, and by that part of the gland, is evident: see Ruysch's plates.

89. *From the foregoing considerations we may deduce the cause of the varieties which obtain in the penis in regard to rigidity or flaccidity. The cause of the rigescence or stiffening is in the main the supply of spirits, by which supply the walls of the cells are not only erected, but also excited to erection by the motion of the spirits, particularly those in the vesiculæ seminales: and this, naturally when there is a superfluity of such: see above. Again also when the arterial, fluid and spirituous blood so much abounds that it can penetrate into the minutest capillaries, and the machine be still so far whole and entire that all efflux into the veins, both the particular veins and the general, is obstructed. Under these circumstances the penis is easily erected by both internal and external causes; but with constantly greater difficulty in proportion to the degree in which these conditions are wanting or are destroyed. Thus the penis grows flaccid from poverty of spirits, from poverty in, or want of, spirituous blood, from the obliteration of the minute capillaries in the cellular substance, from a number of these coalescing into single larger vessels; from the similar coalition of the fibres, which produces a kind of tendinification; from the induration and condensation of the general and particular membrane of the septa, whereby their reaction is lost, and the reaction and action do not correspond: in which case from the very beginning of the act they resist the erection: from the ready escape of the arterial blood into the veins, and from the veins, out of the corpora cavernosa: from failure of desire, namely, when the sharpness of all the sensations and of the imagination likewise perishes. The causes of this are very many in number: as old age, excessive exercise of the generative office, and the rupture of links consequent thereupon; and depending upon the same cause, bluntness of sensation, internal as well as external, arising from the state of the blood; thus from diseases*

and other contingent causes; from poisons; from congenital defect: moreover also from the influence of contrary passions, as anger and sorrow, the former of which hardens the fibres, while the latter contracts them: from the congenital state of the penis itself; not to mention still other causes. For in proportion as the interior causes are defective, the external causes require to be more readily excited. In proportion then to the impotence of the virile member arising from any of these causes, in the same proportion that member contracts, diminishes, shrinks into the scrotum, and conceals itself from view.

90. *Satyriasis or priapism rarely arises from internal causes, but more frequently from external ones. This is shewn by the phenomena of burning desire, which is a kind of obstruction and inflammation that does not permit the arterial blood to escape into the veins; and hence rigidity of the penis is produced. It likewise arises from the sanguineous congestion or obstipation of the veins and cells, which constitutes a second variety of satyriasis: it is declared by many symptoms, but particularly by the state of the glans continued from the urethra; also by a certain spasm of the cells, which cannot regain their pristine situation: and also by the muscles; not indeed by their rigescence, but by their flaccidity, their fibres being relaxed, and unable to react sufficiently: likewise by the pressure on the parts of the anus, whence the arteries and veins come. The erection of the penis after death depends upon similar causes; also especially on the swelling up of the veins and cells, and the hindrance to the return of the blood therefrom, which happens while the corpse lies on its back, for after death the arterial blood is precipitated into the veins.*

CHAPTER IX.

THE MALE SEMEN.

91. BOERHAAVE. "No humor is generated so slowly as the semen; none is retarded by so many passages, or is so much digested by rest. *(Inst. Med.,* n. 650.) The humor which is found in the middle of the testis, in the epididymis, ejaculatory duct, and vesiculæ seminales, when examined in the fresh state, and a little diluted with tepid water, and through a good microscope, is seen to consist of innumerable, minute, oblong, tailed animalcules, like little eels, which swim in the liquid part of the humor. This is true in all cases in man, quadruped, bird, fish, reptile, insect; and always; but only in this situation. If these eels be compared with the carina of the chick as described by Malpighi, in point of size, shape, place of location, and permutations, and with nature's course in the generation of frogs, it will occur as probable, that the animalcules of the male semen contain the rudiments of the future human body; especially when we consider that in case the testicles, or this humor, be wanting, there is always barrenness on the part of the male. *(Ibid.,* n. 651.) The humor of the prostate does not contain any of these animalcules. *(Ibid.,* n. 653.) The semen is quite different in nature from the animal spirits, from volatile oily salt, from hot humor, and from ferment, and is a bland, viscid, inert humor." *(Ibid.,* n. 658.)

92. LEEUWENHOEK. "I opened several gnats, and discovered in their male seed, with the utmost distinctness, a large number of animalcules, which however were much less than those which are seen in the semen of the flea. *(Arcana Naturæ Detecta,* Ed. Nov., Lugd. Bat., 1722, p. 22.) I remember hav-

ing previously described and sent specimens of striæ or fibres, which were lying without any regularity entangled one with the other in the male semen. These however I now entirely reject: since in the male seeds of quadrupeds, birds, fishes, and even insects, I have found animalcules. (p. 27.) In fishes, the masculine seeds are equally large in the smallest individuals as in the larger ones of the same species.—In a minute drop of the semen of a ram I found an incredible multitude of animalcules. —Incomprehensible is the number of animalcules observed in the same semen: some of these creatures are adult, some not so: they succeed each other in oblong troops. (p. 20, 30.) A young ass of a good size in the space of one month readily emitted as much seed as would fill a hat. Leeuwenhoek saw animalcules in the epididymides: he figures the seeds as so many little eels. (*Op.* I., 1, 1722, p. 52.) He observed this more clearly in the seed of frogs, the first shape of them being vermicular. (*Ibid.*, p. 50.) For every female ovulum, the male emits ten thousand of such animalcules, which are contained in its seed. (p. 52.) In the ovum of the flea nothing was to be seen but globules swimming in a watery matter; and the creature becomes at length a worm, and at last an aurelia or chrysalis. (p. 53.) In the vas deferens of the dog there were also animalcula. (*l.* 2, p. 2, 3.) In the cock, a small substance like a particle of sand contained more than fifty thousand animalcules, which were like fresh-water eels. In many places these were grouped and conglomerated so thickly and firmly as to form a kind of opaque body; but in a little time crowd after crowd of them after being thus condensed, again separated. (p. 5.) In the epididymides likewise; which appeared however to have upon them a stratum consisting of numerous very fine globules, and also many plano-oval bodies, which were nearly equal in their circumference to the globules of human blood, to which indeed life may equally be attributed, and this, on the ground of the commotions they undergo." (p. 6.)

93. SCHURIG. "One Langius wrote of these little worms. After it was found that other observers could make up equally good worms with the microscope, it was clear enough that Leeuwenhoek had been grossly deceived by his preconceived

opinion: to this extent at least, that what he looks upon as so many tailed worms, so vastly numerous, are none other than the active portion of the seed agitated within the viscid molecules, and which very gradually consumed, leaves behind it certain motionless fibres, which this author parades as the carcases of worms." (*Spermatologia*, 1720, cap. iv., § 5, p. 165.)

94. To *understand the nature of the seed, we must thoroughly understand the nature of the soul, also of the animal spirit and the blood, for the seed arises evidently and distinctly from them all. The seed is the extract and complement of all; a kind of quintessence which lies most deeply within them. Whatever we term simple, and whatever is the first principle of composite things,—all must pass unclothed and naked into the department of the seed, to make it a substance wherefrom nature can begin anew and live anew. To enable us to form some idea of what the seed is, we must premise certain universals concerning the soul, the animal spirit and the blood; without which we should certainly be hallucinated, and seem to be enlightened with a sort of intellectual light analogous to the ignis fatuus, and playing over unseen things, as it were in a dream.*

95. *As respects the soul, no one doubts that it is a substance, and in fact the simplest, first, innermost and highest substance of its body; consequently the entity or very being of all the perfections of which we form any idea as being below the soul in material entities or things. This substance therefore it is which the other substances in the animate body respect and receive as the chief to which they are subject. Nor is it doubted that the soul and its substance is also the real essence, consequently form, which from its excellence deserves to be called the spiritual and celestial form.* The nature of the celestial and of the spiritual forms will be seen in our Doctrine of Forms. Nothing is more ancient or more frequent with theologians and philosophers than to term souls, and even angels, essences and forms. However it is not enough to call the soul a spirit, and to abstract from spirit all material qualities, which no doubt are to be abstracted: but we

have further to deliver *how* they are to be abstracted, which has to be explained in the Doctrine of Forms. Otherwise, if we merely abstract, and do not demonstrate the abstractions, we shall fall headlong into the thickest darkness, until we positively know not whether the spirit is anything or nothing, inasmuch as nothing is adequate to it which our senses shew us to be anything. *If the soul be an essence, it follows that a nature, power and force lie in it; and if a force, then something that is analogous with activity and modification, or with action and mode: and if with action and mode, also with other accidents and qualities, of which we can only procure an idea from material things. Yet still nothing is said; for substance, essence, form, nature, power, force, mode, accidents and qualities, are words and bare terms, which have nothing in them, unless we have procured by singulars a distinct idea of singulars: in short, by comparison with those things which we know for certain to be in the soul, and which require to be comprehended by thought alone abstractedly from words, although they have to be expressed by words.*

96. *Next succeeds the animal spirit, which is the means of union between the soul and the body; or the middle substance, essence, form, power, force, between the soul and the body; consequently the sharer in both, which must draw its property from the soul on the one hand, and from the purest elements or elementary substances which belong peculiarly to the purer nature or world, on the other.* In this wise this mean or middle term acts, but the place of its action will be shewn when we come to treat of the animal spirit. We must first however speak of the fibres which carry that spirit: and we must also form some idea of the first elements of nature, as they are called. The reader may therefore well imagine the extent of labor required to disentangle one such substance, and draw it out of darkness. But we will try!

97. *Last comes the blood, which partakes of the animal spirit, and at the same time of the elements of nature's lower degree, namely, of the principles of sulphurous and saline substances: and by means of the animal spirit, partakes also of the soul. In this way the blood is the soul's vicegerent in the body, or in the ultimates of the animal world, and may in a certain*

sense be called the bodily soul. But we have already treated of the blood in Part I.

98. *But the determinations of the first substance, essence or soul are called simple fibres, for they are the determiners or parents of the rest. On the other hand, the determinations of the animal spirit, to be represented as composed of simples, are the proper medullary fibres of the brain and nervous fibres of the body. The determinations of the blood however, are the fibres again compounded, or the vessels, arterial and venous. Of these, and of the fibres proper, the motive fibre is composed, which is the last of the fibres considered as determinant.* The reader will see these subjects unfolded in our introduction to Rational Psychology, where we shall endeavor to deliver distinctly the doctrine of forms, of order and degrees, of representations and modifications, and to treat also of the several fibres.

99. *And as it is of importance that we should comprehend what the seed is, and how the soul of the offspring is produced and brought over from the soul of the parent, it is necessary to use words of this kind which are familiar, and give a comprehensible idea of what we have to say: though I am not unaware that in the proper sense they by no means square with the predicates which are or ought to be made, as proper to the soul. Thus if we are to speak of the spirit spiritually, we must not speak of the immaterial materially. Yet as they coincide in the way of representation and analogy, they are at all events instrumental in forming ideas, which when seized in a perceptible manner, cause somewhat to be perceived, which is left to thought alone. The words in common usage signify only bodily and material things, and their modes, qualities and accidents, which if we are not permitted to use on account of the objects to be signified, we shall have no choice remaining but to be quite dumb and utter nothing at all; and even think nothing: inasmuch as what we have to think of, has to be insinuated in the first place through the channel of the senses, and by our imagination and representation of sensible things. Things sensible and imaginative may be represented by vocal formulas, and the mind must afterwards abstract the essences from the latter, which cannot be polished and as it were sublimated by any words short of a continual process of circumlocution and modification. This way of understanding is our natural me-*

thod, for we are instructed a posteriori, *and not immediately* a priori. *For these reasons, in the preface respecting the seed and the composition thereof, we are bound to use words which are not exactly adequate, such as globules, spherules, fluid, membranes, tunics, crusts, figure, space, extension, motion, congregates, volume, and several others, which however, properly speaking, are not suitable to the soul, nor indeed even to the animal spirit, as I shall demonstrate very clearly in my introduction to Rational Psychology. Therefore, for the reasons above given, I hope the reader will not impute this to me as a fault.* No one indeed will do this, unless he be of those who lay it down that nothing is to be said of the spirit, excepting in formulas which are equally obscure and abstract with spirits themselves; in fact in such as nothing suits that involves anything material whatever. At this rate all description must be utterly rejected, and we must remain for ever in occult qualities. Except the unknown and occult be signified by the known, as in the analytic calculus, they continue occult for ever; and the darkness is thickened, not dispersed.

100. *Premising these particulars, we seem now to be somewhat equipped for proceeding to the description of the seed. Now the veriest seed, which is formed first in the testicles, and afterwards in the corpus Highmorianum and the epididymides, and lastly in the ejaculatory vessels or vasa deferentia, consists of globules or most pure spherules, in the inside of which are contained the first, simplest, inmost, highest substances of the animate body; while the animal spirits constitute the surface, which extends around like a crust, and completes the actual globule.* Here we at once find unequal, inequate or inadequate expressions, as globule, spherule, surface, crust, central space; consequently the ideas of space, volume, figure, and so forth. But as I said before, I beg pardon; and will assuredly satisfy the reader in my explanation of forms, orders, and representations. Meantime necessity forces me to use formulas such as the above, to signify the purest and even spiritual things. It is not given to any to speak abstractedly, supereminently, spiritually, but humanly; although it be given to think more purely.

101. *In the very simplest substances, that have the inmost place, and as it were form the nucleus, lies all that which is attri-*

buted to the soul: and in this way it may be said, that it is the soul which occupies the inmost, and by the surface formed by the animal spirit, is so far limited as to represent at the same time a kind of most pure globule or little seed, which constitutes the animal principle and is most perfectly animate. This prime essence, which is the highest and the first in every animal, nowhere occurs so pure, or so gathered and grouped into one unity, as in the testicles, which are formed expressly to draw this essence out from the simple fibres. This is immediately met by the essence of the following degree or the next lower essence, which is what we call the animal spirit. This latter results from the genuine and first compounded fibres, as well as from the blood whose globules have undergone resolution; and straightway it rolls around in circles, and produces a species of surface; which happens in the epididymides, in which the fibres of the whole testicle, the nervous or albugineal coats, and also the purer blood of the spermatic vessels, meet together. And what is begun in the testicles and epididymides is completed in the vasa deferentia; particularly in the more capacious part of the channel of these vessels, and in the whole course of their descent to the vesiculæ seminales. The reader will find the whole process mentioned in the Chapters on the Testicles, Epididymides and Vasa Deferentia. *These most pure globules are the little seeds and the sperms that are fruitful and prolific: and thus it is they that carry the soul and life of the parent, and commence and set up the life of the future offspring. The nature of that first essence, which occupies the inner parts, is too exalted to allow us to give an idea of it simply in familiar words; for it is the substance and formative force, and from the first living point draws the thread, which it afterwards continues onwards to the last point of life.* (Part I., n. 253.) *And simultaneously, and in the present, it seems to have with it and in it the ends, as well the first, as the middle and the last, according to which the causes follow in a foreseen and certain order, to the ultimate effect.* (Part I., n. 260.) *And it represents to itself the state about to be formed as a state formed; yea, and the state formed as about to be formed.* (Part I., n. 261.) *Thus the series of all the contingents for perfecting the work of formation, are with it and in it in a present moment.* (Ibid., n. 263, 265, seqq.) *As soon therefore as ever these properly seminal and spermatic substances come flowing together, as in the epididymides and vasa deferentia, they cannot but at once essay and attempt the work of formation; but in proportion as there is a supply of similar substances, which are requisite for the act and progress of formation. As however these are wanting at this early stage, by means of those which are present, they arrange themselves suitably to the form of the future body in the beginning, into a primitive species of carina, like that seen in the eggs of fowls; that is to say, in determinations and fibres which have the appearance of little eels. These, by reason of their intestine motion and ceaseless activity, represent living animalcules; whereas they are only the first determinations, and the little threads or stamina thus arising, and united to form a series of the kind, from which they seem to design to begin the first web of the future body. And as the parent life and soul lies in every globule, but hitherto indistinct and imperfect as in a rudiment, so these forms are very active indeed, and deceive the most sage enquirers into the belief, that the first origin of the fœtus is taken from them before it begins in the ovulum.* Malpighi is completely in this frame; and Boerhaave also accedes; for these forms insinuate themselves into the senses livelily enough by their eely form and motion, and through the senses into the rational mind: particularly when this view is strengthened by proof derived from the generation of frogs, which are produced from little eel-shaped animalcules; and also when the same is confirmed by the resemblances presented in the incubated egg. And these illusions cannot be dispersed, until we have run attentively through the entire series of formation. For infinite things occur that no reason can explain: for instance, the existence and subsistence of these forms in the vesiculæ seminales, where they abide sometimes for weeks and months, and in the meantime are contiually subject to an incessant motion—both an intestine motion of their parts, and an external motion of the septa. Or take as another point the fact of their getting into the little ova which are impregnated, and which appear to offer no passage or door of entrance. On this account the matter has been more closely examined by the English and other observers, who have used the microscope, and at last it is found, as Langius says, that the animalcules are nothing more than the active portion of the seed agitated within viscid molecules, until being very gradually consumed, it leaves behind it a number of motionless fibrils. And Leeuwenhoek himself, when examining these little eels in the seed of many quadrupeds, birds and fishes, and estimating them at fifty thousand in a space equal to a grain of sand, particularly in the epididymides and vasa deferentia, makes mention at the same time of having seen globules, and even striæ or fibrils. Respecting the globules his words are these: "Lying under the little eels a number of exceedingly fine globules were seen, and many plano-oval forms, nearly equal in compass to our own blood-globules, which may also be said to be alive, judging from the commotion they make." And in another place, in examining the nervous juice which escapes from the minutest fibres, he asserted that little eels creep through those very fibres; but afterwards he confessed his error, and found that what he took for animalcules were nothing more than the continuation and peculiar fibrous cohesion of the nervous humor.

102. *These are the primitive globules, in which lies that vital and natural element that serves for the beginning of formation in the egg. As however they are so very pure, to the last degree agitable, and constantly attempting to begin the work of formation wherever they meet with middle substances, so they require to be again reduced into a more general and a compound form, not very much unlike the first seeds in the vegetable kingdom, which are enclosed in a kind of fruity sheath, and gradually fill up and ripen. For if they were to come out singly and eel-shaped, the work of formation and generation would come to nothing. Now in order to ensure their subsistence and consistence, when the time has come, and they are to be committed to the womb and the ovum, they are carried down into the vesiculæ seminales, whither several species of suitable fluids and elements are conveyed; and in the vesiculæ they are incessantly and distinctly agitated and whirled: by which means these simple forms of the seed are mutually adapted to each other; and again a number arranged under a denser membranous crust, are carried round and round, and new and compound globules arise, which are the proper seminal globules.* This is the second degree of composition; and thus these globules primitively grouped to form others, also suitably located, and surrounded by a more general coat or crust, com-

plete their second dimension. Similar globules are indeed seen by the microscope, for they attain dimensions which begin to be recognizable by that instrument. The means too are present, in a good supply of animal spirits, of the purer elements of the blood, of the unctuous substance from the glands, the watery from the fine cellular tissue; also in the intestine and modificatory motion of all,—a motion for the most part of three degrees, both in the sperms, the fibres and the septa; and also a more general motion of the membranous envelope, which is finely muscular. These causes produce circumgyration of the seminal globules, and determination from cell or loculus to cell; thus perfect commixtion; and when the proper interval has elapsed, and the incessant fostering heat been applied in the meantime, the parts are at last fitted conveniently together, and the intimate essences of nature, the alone determinant and active [essences], are coördinated in forms answering to the purpose they are to serve. These little eels in these cells, during the continuance of this very active and gyratory state, by no means maintain themselves, but separating into primitive globules, they have to pass over without doubt into the genuine forms of the seed. Hence it is that the seed becomes a humor; for that such it is, is manifest from its consistence, density, color, and other visible phenomena, and even from the chemical solutions of it which may be made.

103. *Thus the seed is perfected, and at the same time guarded against all injuries arising from circumstances, as it runs through the urethra: moreover accommodated to flow into the womb: also when it breaks out of the vesiculæ seminales, it is besprinkled with a new humor, namely, that of the prostates, and enveloped therein, to prevent it from being let out of its membranes prematurely, and bursting forth, and making the work of generation null and void. And lastly, it is irrigated and inviscated with another humor in large abundance, which besides giving it quantity also supplies it with determination, whereby it is emitted into its ground or soil, not spontaneously, but with the force, and in the manner, of ejaculation.* Respecting all these subjects, see what we said above on the Prostates and the Urethra.

104. *In order that this seminal material, thus perfected and transmitted into the womb, may commence from its first principles*

to form the new life and the diminutive body, it is necessary that it be again resolved so far as regards this second composition, in this its ground or womb, as the seeds of the vegetable kingdom are in their field: that so the naked seminal globules like a vital aura may fly out through the Fallopian tubes into the ovula seated in the ovaries. The fact of this resolution is to be shewn in what follows. That it does take place is confirmed by the parallel case of vegetable seeds, which are enclosed in their own proper membrane, and some kinds of them are again surrounded with a particular dense substance, as the seeds of grapes, berries and a thousand others. The reason of their being covered over with a denser coat, is that they pass into a hard soil or genital field. For all things correspond. According to the nature of the ground, soil or womb, a particular habit is assumed; of opening out and unlocking themselves, quicker or slower, according to the causes presented in the field or womb. The globules thus unlocked or opened are analogous to a kind of vital aura, being to the last degree light and volatile, and in mobility equalling the very auras. So also they penetrate the purest pores of the ovules as a sieve-like tunic. It would not be so if they were eel-shaped, and had to creep up to, and through them. *The ovules seated in these ovaries, are formed in such a manner, as to offer them (the globules) not only an opportunity of insinuating themselves, but also to afford them a pouch the most fitting, which in fact will admit a goodly number of them to be heaped and grouped together, and not only this, but also will enable them to procure for their first nutrition the most suitable substances. Thus this first, inmost, highest animal essence, in which the soul lies, and which the soul is, commences its web of formation suitably to its nature, and first of all conceives and excludes a cerebrum in the purest type, or such as it is in the depths of the cortical substance; and ultimately from many such, by means of the simple fibres, extends itself to form the carina; and so forth.*

105. This however is not the proper place to shew the mode in which the work of formation is carried on; but only the fact, that all things begin from the first globule or as it were ovule, and that the ratio or manner of the whole formation, lies in the formative substance, which when thus left to itself, can-

not but regard the *formed* as *to be formed,* and the *to be formed* as *formed;* and all the ultimate effects as simultaneously present that in the first representation and view are presented as such. This soul residing in the principles of natural things, looks upon the whole of nature as lying in a manner below it, like one who looks upon the fields from a mirror on a tower. Universal nature is but the instrumental cause which ministers to life, and necessarily concurs to all its decisions; the whole physical world being completely subject to the moral and spiritual worlds: and so nothing can be in the way to prevent the soul from passing most freely to ends, which are the only things that it regards. *The soul which resides in the principles of natural things, and keeps all nature underneath it, to serve life as an instrumental cause, of its own nature so disposes all things, that effects correspond to ends, to which effects the soul passes through the innermost penetralia of the sciences and arts. Thus she puts together the texture of the body, which to our mind's eye is full of miracles; while the soul, of her own nature, produces similar and still other things to infinity, without any previous knowledge of the sciences to aid her.* But these are psychological subjects. *If such be the formation, composition, and form of the seed, we can then explain the several facts that occur in the genital organs, as also those of conception and generation; moreover those that furnish controversies to the world on the subjects of the formation of the fœtus from the seed of the parent, of the evolution of the same in the mother's womb, and of the propagation of the soul in the offspring. For on these principles there is nothing but what is explicable and intelligible.* But these particulars are brought forward respecting the seed only in the way of preface or first essay: the several points will require to be further proved: in that case should anything, when duly considered, not fall in with facts, it will be the surest proof of hallucination; otherwise it may be assumed, if not as demonstrated, still as confirmed by all those proofs and phenomena that are at present extant. If so, then this theorem is in the last term of the probable, and in the first term of the true. We cannot get by any means into the veriest truths unless we be instructed, not *a posteriori,* but *a priori.* That alone which flows from the Divine Mouth is truth. We shall therefore deduce in the sequel, when we come to treat of the soul, how anatomical and at the same time how physical proofs, and lastly how philosophical conjectures, coincide here, and joining their forces, fight under the flag of our induction. In that place we shall again take up the present argument, to confirm *a posteriori* those matters that conduce to our principles.

PART II.

THE FEMALE GENERATIVE ORGANS.

CHAPTER I.

THE EXTERNAL GENERATIVE ORGANS OF THE FEMALE.

106. HEISTER. "The external parts, or those which may be seen without dissection, . . . are these: the pudendum or vulva, in the midst of which is a fissure, and the orifice of the vagina; at the lower part is the frænulum, and the perinæum. The two labia, and the mons veneris, which is tumid from the fat that lies under it, and is beset with hairs. . . . The clitoris, or as some call it, the mentula muliebris or female penis. The clitoris is situated in the upper part of the fissure, and lies for the most part almost entirely out of the way of the stimulus of venery, concealed under its skin or prepuce. Its ordinary shape and size are about that of the point of the little finger; sometimes however it attains extraordinary dimensions, so as even to equal the male penis; but still it has no urethra running through it, and is imperforate. It has a glans or apex, which also is not perforated; but nevertheless is smeared with a fetid humor, like the corona of the penis. The prepuce covering the glans of the clitoris is continuous with the cutis of the pudendum, and is furnished with nervous papillæ, which render it very sensile. It has also a ligament, which connects it to the ossa pubis, almost as is the case with the penis. It has two crura, thrice as long as the clitoris itself, and coming from the ossa pubis. It has two corpora cavernosa, with a middle septum or partition, in this respect again resembling the male penis; these crura make up the bulk of it, . . . and are surrounded by a nervous membrane. Its erector muscles are two in number, as in the penis: they arise from the ossa ischii, and are inserted into the corpora cavernosa. Its arteries and veins come from the hypo-

gastric and pudendal vessels. Its nerves arise from the sacral nerves; and considerable branches of them run along the back of the clitoris; whence its great sensitiveness.

107. "The nymphæ are two membranous parts, situated one for the most part on each side of the fissure. They are reddish, cavernous or spongoid in structure, and somewhat resemble the wattles under a cock's throat. Their size varies, being sometimes larger, sometimes smaller; and they are continuous with the prepuce of the clitoris, and connected to the inner side of the labia. They are abundantly covered with nervous papillæ, and hence are extremely sensile. They also have little glands, which secrete a sebaceous substance. Besides the offices of titillation, . . . they are also of use in directing the stream of urine. . . .

108. "The orifice of the vagina, or external os uteri, is surrounded with a cavernous substance, which swells up during venery. In virgins it is comparatively narrow: in other persons, and particularly those who have borne many children, it is more ample: always however it is narrower than the rest of the vagina.

109. "The hymen is a membrane which varies in shape; sometimes it is circular; sometimes semilunar; and sometimes it takes a different form from either of these. It bars the vagina in virgins, and is always present in very young subjects. (Some deny this latter position, but Heister has constantly found the hymen at any early age: many other authors are agreed with him in this experience. See his *Not.* 30*.) In little girls the hymen has a small aperture leading into the vagina: in adults however the aperture is generally larger. The hymen is lacerated on the first coition (unless indeed it has been torn before), and frequently bleeds on that occasion. The carunculæ myrtiformes arise from the laceration of the hymen; and are observed to the number of two, three, or four, in the place where the hymen formerly was. They are not present in the infant.

110. "The urethra or meatus urinatorius lies straight beneath the clitoris, its site being marked by a little eminence. It is about two finger-breadths in length. It is of larger calibre in women than in men: at the end however it is narrower than

in the rest of its extent, and it admits of great dilatation. There are certain little ducts which open into it, and which excrete a mucous liquid, as a protection against the acrimony of the urine. The origin of these ducts, as of the corresponding ones in the male urethra, is not yet known with certainty. Then also there are the lacunæ Graafii, small orifices visible about the urethra. These are ducts that excrete a humor for lubricating the vagina, and they arise from the glandular substance of the urethra." (*Comp. Anat.*, n. 234.)

111. WINSLOW. "*Pubes.*—The pubes is that broad eminence which lies exteriorly at the lower part of the hypogastrium, between the groins, and on which hairs grow at a certain age, which are called in Latin pubes, and are almost similar to those found under the axillæ. The prominence of the part is owing to a particular thickening of the adipose membrane that covers the forepart of the ossa pubis, and some small portions of the neighboring muscles.

112. "*Sinus and labia or alæ.*—The longitudinal cavity which reaches from the middle and lower part of the pubes, to within an inch of the anus, was by the ancients termed the sinus; and they called the lateral part of that cavity, the alæ. . . . . The places where these alæ are joined above and below, are termed commissures; and may likewise be called the extremities or angles of the sinus. The alæ are more prominent and thicker above than below, and lie nearer each other below than above. They are chiefly composed of the skin, spongy cellular substance and fat. The exterior skin is a continuation of that of the pubes and groins. It is more or less even, and furnished with a number of glandular granules, from which a whitish ceruminous matter may be expressed; and after a certain age it is likewise covered in the same manner with the pubes. The inner side of the alæ is something like the red portion of the lips of the mouth; and it is distinguished everywhere from the external part by a kind of line, in the same manner as the red portion of the lips from the rest of the skin; being likewise thinner and smoother than the outward skin. A great number of pores are observable in it, and also numerous glandular corpuscles are embedded in its substance, and furnish

a liquor more or less sebaceous. These glands are larger near the edges than in the other parts.

113. "*Lacunæ.*—Near the inner edge of the inner surfaces of the alæ, on each side of the orifice of the canal of the uterus, we find a small foramen more visible than the rest. These two foramina are termed lacunæ. They communicate by two small ducts with two follicular bodies lying in the substance of the alæ, and which may be looked upon as little prostates answering to the prostate glands in man. When compressed they discharge a viscid liquor. Above the superior commissure, a thin flat ligament runs down from the small ramus of each os pubis, which penetrates the fat in the substance of each ala, and is there lost insensibly near the edge. These may be looked upon as the suspensory ligaments of the alæ. The inferior commissure of the alæ is very thin, like a membranous ligament, and together with the neighboring parts of the inner sides, it forms a depression, termed the navicular or scaphoid fossa. The planes of these ligaments are nearly parallel. The space between the inferior commissure of the labia and the anus, is termed the perinæum, and is about a large finger-breadth in length.

114. "The other external parts are situated and concealed in the sinus, and hid by the labia. Directly under the superior commissure of the labia lies the clitoris, with its covering or prepuce. A little lower down is the orifice of the urethra. Below that again is the orifice of the great canal of the uterus. The circumference of this canal is bordered either by a membranous circle, called the hymen, or by fleshy portions, termed the carunculæ myrtiformes. On each side of the clitoris begins a very prominent fold, like a crest, which runs down obliquely on each side of the orifice of the urethra. These two folds are termed nymphæ, and they might likewise be named the crests of the clitoris. On each side of the great orifice lies the small prostatic foramen already described.

115. "*Clitoris.*—The clitoris appears at first sight like a small imperforate glans. Above and laterally it is covered by a kind of prepuce, formed of a peculiar fold of a portion of the inner side of the alæ. This fold appears to be glandular, and to discharge a certain moisture. On the inside it is gra-

nular. By dissection we discover in the clitoris a trunk or stem and two branches, almost as in the penis; the whole made up of a spongy substance, and of very elastic coats, but without any urethra. This substance may be inflated either by air or by anatomical injections into the artery, &c. The trunk is divided into two lateral parts by a middle septum, from the bifurcation to the glans, where it is insensibly lost.

116. "The bifurcation of the trunk is on the edge of the cartilaginous arch of the ossa pubis. The branches, which also resemble the roots of the corpora cavernosa, are attached to the edge of the inferior rami of these bones, and extend interiorly over the small ramus of the ischium, where they terminate by degrees; although one portion of the membranous tube on each side sometimes reaches to the tuberosity of the ischium. The trunk of the clitoris is sustained by a suspensory ligament attached to the symphysis of the ossa pubis, and which encloses this trunk in its duplicature, nearly as in the other sex. Four muscles or fasciculi of fleshy fibres are attached to the trunk of the clitoris, two on each side. One of them runs down along the neighboring corpus cavernosum, covers it in front, and is attached by a tendinous or aponeurotic portion, partly to the extremity of the corpus cavernosum, and partly to the tuberosity of the ischium lower down. These two muscles are called erectors. . . . The other muscle on each side lies immediately below the former, and runs down on the side of the urethra and of the vagina uteri, enlarging as it goes, all the way to the anus; where it terminates partly, almost like that which is termed the accelerator in the male subject. These two muscles of the two sides surround very closely the lateral parts of the urethra, and of the vagina. They expand considerably as they descend, and overspread the lower and lateral parts of the great orifice; for which reason several anatomists have looked upon them as sphincters or muscular girths. These four muscles, and especially the two latter, are often almost covered with fat. The blood-vessels of the clitoris come chiefly from the hypogastrics. The nerves arise from the second and third pairs of sacral nerves, through which they communicate with the inferior mesenteric plexus, and with the great sympathetic nerves.

117. "*Nymphæ.*—The nymphæ, or crests of the clitoris,

or as they may likewise be termed, alæ minores sive internæ, are two very prominent folds of the inner skin of the great or external alæ, reaching from the prepuce of the clitoris to the two sides of the great canal of the uterus. They begin very narrow and pointed, and increasing in breadth in their course downwards, they are again contracted at their lower extremity. They are of a spongy substance, intermixed with glands, several of which are perceptible to the naked eye. They are situated obliquely, their upper extremities approximating, while the lower divaricate. In married women they are more or less flaccid or *flétries*. . . .

118. "*Urethra.*—The orifice of the urethra lies in the interval between the nymphæ, below the glans of the clitoris. The sides of this orifice are a little prominent and wrinkled, and perforated by several small lacunæ, from which a more or less viscid or mucilaginous liquor may be squeezed. In time of pregnancy, this orifice is sometimes drawn a little inward. The body of the urethra is a spongy canal of the same structure as in males, but very short, situated directly under the trunk of the clitoris, and above the great canal of the uterus, adhering by membranous filaments to each of these canals between which it lies. It passes under the cartilaginous arch of the ossa pubis, and terminates by an oblique opening in the neck of the bladder; being bent a little downwards between its two extremities. The internal membrane of the urethra is a little plaited, and perforated by small holes, which communicate with follicles lying hid in its substance, as in the man. If we blow into one of these holes, we observe a small canal to be inflated, which runs from without inwards, and terminates in some places in a kind of ampulla, on compressing which, a viscid liquor is discharged. The continuation of this membrane, which lines the neck of the bladder, forms likewise several rugæ more or less regular, but that which lines the cavity of the bladder is wrinkled in an irregular manner when the bladder is empty.

119. "*The canal of the uterus.*—The great canal, formerly called the neck of the uterus, is situated below the urethra, and above the extremity of the rectum, a little obliquely, being more raised on the inner and back part, than on the outer and fore part. Its inner or posterior extremity joins the extre-

mity of the body of the uterus, and surrounds its orifice much in the same manner as the duodenum surrounds the pylorus, or as the ilium is surrounded by the cæcum and colon. The anterior extremity forms the great orifice, which lies under that of the urethra, and above the fossula of the inferior commissure of the labia. The body of the canal is chiefly made up of a spongy substance, interwoven with quantities of blood-vessels. It is commonly longer and narrower in virgins than in married women. Its inner or concave surface has several transverse rugæ, and is covered by a particular membrane. The rugæ are formed by oblong narrow eminences, incurvated like portions of arches, placed very near each other, and arranged in such a manner as to divide the concavity of the canal into an upper and lower side. By the union of the extremities of the upper and lower arches of rugæ, a kind of raphe or irregular suture is formed on the right and left sides. Both these arches are sometimes intersected in the middle, and so form two half-arches; but in this there is some variety. In general, these arches are very considerable in young persons, and in virgins. They become more superficial in married women, and are effaced in time of parturition. The inner or posterior extremity of this great canal surrounds the orifice of the uterus a little obliquely, in such a manner, that the upper side of the canal lies very near the orifice, and the lower side at a greater distance from it; this causes the extremity of the uterus to appear to advance more into the canal on the lower, than on the upper part.

120. "*Circulus membranosus.*—The external or anterior extremity of the great canal in virgins, and especially before the first eruption of the menses, is commonly bordered by a membranous fold, more or less circular, smooth, and large; sometimes it is semilunar, and in some subjects it leaves but a very small opening, in others a larger opening, but ordinarily renders the external orifice narrower than the rest of the cavity. This fold is the hymen, and it is formed by the union of the internal membrane of the great canal, with the membrane or skin on the inside of the labia. It represents a membranous circle of different breadths, and sometimes uneven.

121. "*Carunculæ.*—This membranous circle is commonly ruptured after the consummation of marriage. It is effaced in

delivery; and afterwards only some irregular portions of it remain, which from their supposed resemblance to myrtle leaves, have been termed carunculæ myrtiformes. This circle may likewise suffer by too great a flux of the menses, by accidents, imprudence, or levity.

122. "*Plexus retiformis.*—Each side of the anterior portion of the vagina is covered exteriorly by a thin broad cavernous and vascular plexus, called the plexus retiformis of that canal. These two planes run down on each side of the clitoris behind the nymphæ, and likewise cover the urethra like a collar, before they are spread on the great canal. This plexus is closely united to the inside of the muscular girths commonly taken for accelerators or constrictors, lying between these portions and the lateral parts of the urethra and of the great canal. The plexus may be inflated by air like a flaccid spleen, or like the spongy substance of the clitoris, with which it seems to communicate. On this account the lateral portions of this reticular plexus have been named the internal crura of the clitoris. It is a kind of rete mirabile of vessels, which come chiefly from the hypogastrics." (*Exp. Anat., Tr. de la Teste.*, n. 624—657.)

123. SCHURIG.* *Perinæum.*—Cases of ruptured perinæum arising from delivery; and from the presence of wind also concerning the extraction of calculus through the perinæum. (*Embryologia*, Dresd. and Lips. 1732; 4, vi. 5, § 8, p. 561 ss. *Muliebria*, 1729; p. 37 ss. 42, 45.)

124. *Vulva.*—Extraordinary size of the vulva: fire issuing from the vulva: (?) double vulva in lizards, crabs, and rays. Vaticination by the vulva. Itching, inflammation, fungus, of the vulva. The passage of fæces through the vulva. The vulva is sometimes closed. (*Muliebria*, p. 49—66.)

125. *Clitoris.*—Ossification of the clitoris: its size is very various: it is the cause of pleasure: it is sometimes large in children, and in adults, and even hinders the person from walking. Is it a proof of hermaphroditism? Of the tribades, or women "*quæ subigunt alias.*" (*Ib.*, p. 80—95.)

* The quotations from Schurig are mere notes, and not literal citations. Swedenborg has undoubtedly made great use of the curious treatises of Schurig, in many of his inductions. There are few points of novelty in the latter, excepting, what we always note in Swedenborg, the combination of the ideas into a whole.—*Tr.*

126. *Nymphæ.*—The nymphæ are little membranes generally of a fleshy texture, always acuminated below, and broad above. In virgins they are a kind of ligament, and in a manner constrict and bar the lower part of the external orifice: above they are connected at an acute angle, where they form a corrugated membranous process, which is like a prepuce investing the clitoris. In young subjects they are soft; afterwards they are indurated, and become almost cartilaginous: this is according to Riolan. They vary in size and shape. On opening the bodies of females after childbirth, he found them nearly obliterated. The circumcision of women among the Persians and Ethiopians consists in the amputation of the nymphæ. *(Ib.,* p. 129 ss.)

127. *Carunculæ myrtiformes.*—These bodies are single in virgins; but there are several of them, sometimes as many as four, in married women: and there are an equal number of fleshy membranes to attach them. But they appear to consist of the rugæ of the vagina: they are said to arise from the breaking up of the hymen: they bar the orifice of the cervix.

128. *Prostates.*—De Graaf says that the prostates of women have an irritant function; the humor which they secrete being pituito-serous; which by its acrimony and saltness renders the women more salacious. Some authors erroneously regard it as the female seed. The seat of gonorrhæa in the female is in these prostate glands. *(Ib.,* p. 164.)

129. *The hymen.*—Some writers describe it as a membrane of a membranous substance, while others variously reckon it nervous, fleshy, ligamentary, venous, retiform; some also say that it is thick, and some that it is thin. Some say that it quite precludes the passage; but this is denied by others. The hymen consists of the carunculæ.

130. *The vagina uteri* lies between the bladder and the rectum. In virgins the orifice is much narrower than the vagina itself. It is oblong, contorted, and a little oblique. When coitus is not transacted it is flaccid and collapsed; on the other hand, during menstruation, copulation, and parturition, it is four inches long, and half an inch broad. After delivery it is larger than before. In the venereal orgasm it responds to the virile member, which it closely embraces. In the lower part of

it we find a great number of little pores; out of these the humor gushes attended with great pleasure. The glands are in clusters like grapes. In one subject a fleshy excrescence was noticed. Disruption of the vagina towards the rectum has also been known to occur. *(Ib.*, p. 169, 211.)

131. BOERHAAVE. "The blood-vessels, lymphatics, and the adipose and nervous vessels, the membranes also and the fibres, are far more lax in women than in men. Hence all their cavities, cells, vessels, are more readily filled, and their humors aggregated: so that women have always a very great thickness of the cellular and adipose membrane." *(Inst. Med.*, n. 661.)

132. MORGAGNI. "Besides the lacunæ which belong to the glandular substance of the urethra, . . . I have not seldom found that the other particular foramina (which are disposed about the beginning of the vagina, immediately in front of the hymen, or where the hymen has been torn, in front of the carunculæ myrtiformes) are more than two in number. I have counted two, three, and even sometimes as many as four of them on each side. And it sometimes happens that one finds one lower down, almost in the median line. Not long ago I saw one foramen lurking almost under the glans of the clitoris, and a short duct placed transversely terminated in it. . . . Sometimes the foramina are exposed to view, but sometimes again a portion of them are covered, the carunculæ falling forwards upon them. . . . There was no difficulty in introducing even a largish bristle into each of them. . . . Some of them run upwards, some pass downwards. . . . I have sometimes seen one on each side which belonged to a roundish gland of no mean size. . . . Within the gland, in the place of a receptacle and cavity or pelvis, there were several vesicles or cells, the one opening into the other, which were full of a limpid humor. These were the beginnings of the duct before mentioned. *(Advers. Anat.* I., n. 31, p. 42, 43.) Horne and De Graaf . . . stated that a pituitous serum may be got from the little foramina in the cervix of the uterus. Others on the contrary affirm . . . that a glutinous and viscid juice issues from the same pores. . . . Whether I dissected the bodies of women, or of cows, I met with . . . the same appearance of tenacious and limpid mucus, a little however inclining to a yellow hue, in the top of the vagina, and also in the cervix.

. . . I also observed certain vesicles of various sizes, all however roundish, or oval, covering in close groups almost the whole internal surface of the cervix." These Morgagni affirms to be sources of humor or fluid. Some persons, he remarks, consider them as hydatids, but this he questions. "This sluggish and glutinous mucus, . . . during the latter months of pregnancy, becomes much more inspissated and dense, . . . and furnishes a most opportune defence against the invasion of the air."* *(Ibid.,* n. 33, p. 45.)

133. "*The nymphæ.*—Great varieties occur in these parts. They have sebaceous glands. (*Ind. Rer.*, art. Nymphæ.) There are very great numbers of glands over the whole of both surfaces of the labia, if you only except their external border and top portions. . . . These glands are small, . . . and often in the centre of each a little orifice is discernible, and a whitish and very delicate thread of the same hue as the glands, may be observed issuing from it." These Morgagni considers as sebaceous glands. (*Ibid.*, n. 11, p. 8.)

134. "*The clitoris.*—There is a suspensory ligament which stands in the same relation to the clitoris as that which its proper suspensory ligament occupies to the penis. (*Ibid.*, n. 18, p. 20.) In cows, dogs, sheep . . . the clitoris is set in the inferior extremity of the parts." . . . (*Ibid.*, IV., an. 23, p. 42.)

135. "*The hymen* reaches all the way to the meatus urinarius, although in proportion as it mounts, it becomes thinner and finer, so that . . . it rather appears to resemble a crescent moon. . . . . From the upper and interior part of the vagina, following its longitudinal direction, and tending towards the hymen, there descend two, or three, or sometimes more, considerable columns, or rather corrugations of the doubled inward substance of the vagina, and these columns leave a considerable space between them. The nearer they approach to the hymen, the thicker they become; and for this reason, the vagina is narrower towards its lower part. The extreme bases of the columns, both from their shape and size, may probably be taken by many

* Swedenborg here adds, as Morgagni's opinion, that the use of the mucus from these foramina is to "conglutinate the vagina uteri during gestation." This we cannot find in Morgagni.—*Tr.*

anatomists for the carunculæ myrtiformes; for we know that many have affirmed the hymen to be nothing else than these carunculæ mutually bound to each other, and grouped together, by membranes. In this assertion however they are mistaken." (*Ibid.*, IV., anim. xxiii., p. 43, 44.)

INDUCTION.

136. The *parts of the woman answer to those of the male, just as though they had first been naturally united, and afterwards naturally separated; and when united they resembled only a single body, but when separated, two bodies. The female parts, therefore, are fashioned entirely for the reception of the male, for conjunction of fitness with them; for unition, for the advancement of acts of coöperation; and finally, for the gestation or carrying of the fœtus, and for bringing the latter into the world by birth. Added to these qualifications in the parts, there are bland impulsive causes, both in the body, and in the disposition or animus, as well as in the soul itself; yet with a difference like that which subsists between active force and passive force, and between essence and form, which latter, (viz., passive force and form) are also naturally reactive. From these two principles mutually conjoined there results a unity or* a one, *capable of intrinsic completeness, and whence at length a third thing is produced.* But to come to the several points. In general it is a constant fact, that in order for any perfect thing to exist, there must be an agent and a patient; thus, as everything in nature aspires to self-perfection, that which is active has an appetence for the passive, and *vice versâ*. But on these subjects we refer the reader to our introduction.*

137. *The female parts so correspond to the male, that during the time of carnal copulation they resemble one and not two bodies, and so far coalesce that in the moment of orgasm the two are scarcely aware that they are not one: moreover nature at the*

* Probably Swedenborg here alludes to his Introduction to Rational Psychology in the *Economy of the Animal Kingdom.—Tr.*

same time endeavors actually to unite them both. Love, and particularly sexual love, is a most close union of wills and dispositions, and indeed a union so great, that the life of the one is in all ways communicated to the life of the other, so that in the height of the ecstacy, whatever is peculiar and proper to either is thought to be the other's reciprocally. We have shewed above that at that moment the soul strives to go forth out of its body, and as it were to found a colony; for the universal simple fibre, which is a determination of the purest essence, and which in a substantial sense is the soul, opens up in all directions, and endeavors to transfer itself elsewhere. Provision however has been made, that none of the soul shall emigrate but that which is entrusted to the testicles: see above. Were it not for this provision, the ardour is such, that the whole of the vitality and the entire animal principle would emigrate; as we see from the several phenomena of the act, in which every one forgets himself, and believes that he is the other's. The communion of dispositions between the two parties cannot be better shewn than by this very goad which is upon both, and by their actual copulation. At this time then, neither is his own, but the other's; and while this state is reciprocal, the property or peculiarity of each gives rise to that community of which at such a time it is difficult to have a distinct perception. *The union of dispositions is represented physically in the union of the bodies, and of the members which specially serve and help the process. But in order to a perfect union, the whole series of union from first to last must be formed, so that all the parts shall plead one cause. Thus the soul and its representation; the mind, and its will's desire; the disposition or animus and its passion; and the body and its pleasure,—all must concur, and must complete the series; which although in itself successive, must be simultaneous; so that whether representation, or desire, or passion, or pleasure is struck out, the communion, and consequently the coöperation, must cease at once.*

138. *That the female parts are exactly fashioned for the reception of the male, is clear from the several forms and types of the female receptacle, from the surface of the fissure to the inmost cavity of the womb. Moreover, the dimensions of the two correspond; or, if there be any difference, the female parts are adept in enlarging and in constricting, and by their very nature*

open, then seize, embrace, and fit closely to and with. This is the end of the fissure at the orifice, of the nymphæ, carunculæ, the folds of the vagina uteri ; and also of the penis-like clitoris itself. There are so to speak three doors to the womb. The first is the fissure, which in the first instance receives, and closely seizes, and embraces the bottom or root of, the male organ. The nymphæ constitute the second gate, and by their union produce another almost similar fissure; and they further seize, and shut upon, the male organ. The vagina itself is the third gate, the orifice of which is comparatively narrow, but the tube more ample. In the years of virginity this gate, sacred as it were, is closed by the hymen, and afterwards in other years it is guarded by the carunculæ myrtiformes. The vagina itself is of such a make, that it allows itself to be expanded to the size of the organ which comes into it, but so that at the same time it grasps that organ closely through its whole extent; the wrinkles and folds that appear upon it are of a peculiar spiral kind, and this allows them the property of expansion, though they tend continually to a diminution of their calibre. The like provision is ensured in the orifice by the coöperation of the carunculæ myrtiformes, which have a kind of muscular tissue wrought into their fabric, as well as into the orifice itself; for the principal importance here is attached to the just application of the orifice. Moreover, there is the cavernous texture of the inner surface of the vagina, which renders its expansion easy. The clitoris too coöperates, for with the orifice of the vagina, the clitoris makes size in every way, as appears from the application of the muscular fibre: the accelerator muscles also correspond to the accelerators in the male organs, and by their fibres embrace the clitoris, being applied moreover to the orifice of the vagina. Thus when that orifice is expanded, the muscles are relaxed, and room is given for the clitoris to expand. We indicated above, in treating of the penis, that the provision which limits the expansion of the clitoris, and constricts it, thus preventing it from swelling beyond the due size, is found in the muscular fibre. The same may be remarked with respect to the erector muscles of the clitoris, as well as regarding its suspensory ligament. Thus the opening and expansion of the orifice of the vagina lends the clitoris the opportunity of swelling up, and in this way the clitoris contributes to the application of the vagina, or rather of the orifice of the vagina, as well as to that of the nymphæ, to the member of the other sex. The nymphæ likewise, together with the fossa navicularis or scapha, contribute their part, to the capacity not only of expansion but also extension in all the rest; for these nymphæ are folds and reflexions of the same membrane, though united and as it were coalescent; notwithstanding which they relax during the moments of the greatest expansion and extension, unfold or *explane*, fit all the parts to the process, and at the same time tend constantly to assume a lesser size. In this manner all the parts are fashioned for reception, and for fitting to their mutual parts.

139. *Nor do they conspire and strive together only for reception of, and fitting to, their mutuals, but they aim at absolute union. For the latter end a viscid moisture, threefold in kind, is excreted in abundance, and thereafter flows around and about the parts in every direction; and not merely smears and fills in all those spaces where the parts are not in close apposition, but also endeavors at the critical moment to unite the one thoroughly to the other.* Innumerable ducts and pores are visible from which this humor oozes, namely, in the vagina, on its internal surface, as indeed Morgagni shews; moreover, close under the carunculæ myrtiformes, these being larger than the former, and styled lacunæ, of which lacunæ Morgagni enumerates as many as four. The ducts terminate in follicles, and also in the female prostate glands; moreover in the urethra, in the cavity of which, in the female as well as the male, numerous orifices are seen which discharge a viscid humor. Besides these, the whole cavity, the nymphæ, labia, and the other parts, are beset with glandular granules; all which sources provide a copious afflux of moisture, which runs around the male organ, and causes a union so close that not the smallest portion of air can penetrate. This humor is not a seminal but a conjunctive humor, and it mingles in no sense with the male semen. It seems to be of three kinds. That which comes from the vagina and urethra, and perhaps from other sources, is a thin glutinous serum, and appears to be expressed from the cells, follicles, or from the cavernous part of the clitoris, the external surface of the vagina, and also of the urethra. The second sort is a little thicker, and comes in fact

from the prostates so called through the lacunæ. Lastly, there is the sebaceous and ceruminous matter from the glands. These three appear therefore to differ in grossness and density; but all the species are of that nature that they are openly viscid or sticky, and binding like glue; and at the same time they plainly have the effect, of not only filling in empty cavities, but also of aiming to glue the active viscera to the passive during copulation; although matters are so arranged that this result is never attained.* This follows unavoidably from the very nature of sexual love, whose criterion and attribute it is, that perfect union and communion shall exist. This induces necessity on even the physical organs, which aim by correspondence to produce similar union. The effect is known from the cause, and the cause from the effect. The viscidness of the humor we are describing, is plain enough; so also the viscidness of the human semen, which has almost the appearance of a kind of glue where it dries upon the linen. This property however does not prevent the humor from lubricating the passage, and serving to facilitate the business of the act of copulation. Whatever nature produces, involves not simply one end, but many intermediate ends at the same time, and thus, together with the primary use, engenders multitudes of secondary uses. And so, the humor of the urethra is not only for securing that passage against the acrimony of the urine, but also for increasing the mass of humor in the generative moments; for where it flows out during the act, it adds itself to the rest, and appears to be of a similar nature with them.

140. *These same organs, and some others, together with the moisture which oozes from them, foment and excite the passion, pleasure and lust, carrying them to the last effect, goad and fire of which they are thus capable, so that the end follows of necessity. There is an entire series of excitements and allurements. The excitements are divisible into innermost, middle, and outermost; or into primary, posterior, and ultimate: and they require to concur simultaneously to the act, and in the act. The innermost or intimate excitements are those which have their seat in*

* Let the reader mark with reverence the amazing logic of creation, in which the very properties of the fluids carry the same life of love and union that the conjugal relation exhibits on the larger scale.—*Tr.*

the soul and mind; whence our nature is prone to the propagation of offspring: this proclivity is called forth by its own peculiar cause. The middle incentives are those which kindle the animal mind, and produce the blaze of passion; they arise from the allurements of the senses, from sight, hearing, touch; and consequently from the imaginative power, which derives its existence from sensation. The last incentive is the actual contact in the crisis; and occurs when the sensitive papillæ that are seated upon the labia, alæ, nymphæ, and vagina, are roused and stimulated by the continued touch and attrition both of the soft smooth penis, and of the sharp and salty humor which flows between the two. Upon comparing these excitements with the corresponding ones in the male sex, it is plain that females are more prone and vehement in the paroxysm of love, and have less command of themselves, than men. The incentives which exist to venery, and the amount and quality of the same, enable us to conclude very safely as to the degrees and differences of passion. In this place I canvass only those excitements that arise from mere touch, and which are external; for the internal correspond so completely to the external, that no doubtfulness arises in our reasoning from the latter to the former. Now in the female, the external excitements comprise all those papillary, nay, and all those glandular forms which are visible; namely, in the nymphæ, and throughout in the cavity of the vulva, and which anatomy shews to be of a papillary* nature. Then there is the internal tunic, tender and soft, and composed of mere [nervous] fibres; and which is more sensitive than the cutis, which latter is covered by the epidermis. The prepuce of the clitoris, which is continuous with the nymphæ, is also provided most liberally with papillæ. The vagina again is nervous and papillary: see Heister. Moreover, a host of nerves, arising both from two pairs of the sacral nerves, and from the inferior mesenteric plexus, and consequently from branches of the intercostal nerve and par vagum; that is to say, of the sympathetic nerves; and which mounting the clitoris, almost entirely constitute what is termed its nervous coat, and even its cells, prove the same thing. See what we said before respecting the corpus cavernosum

* That is to say, to be, like the papillæ of the skin, organs of touch.—*Tr.*

of the penis. The whole therefore of that little penis, the clitoris, is constructed with an express view to the excitement of venery; and together with the corpus cavernosum urethræ, and with the corpus cellulosum of the external surface of the vagina, and with the nymphæ and perhaps with the carunculæ myrtiformes, swells up; and still more extensively when the vagina is drawn open: see above. The result of all which is, that a sharper sense of touch is engendered in the papillæ and the several organs. The glans of the clitoris is the main organ of excitation; it is applied to the hairy portion of the pubes of the male, and according to the degree of its extension, is excited to the lust of the moment: indeed this glans is applied to the outermost part of the man, at the same time that the glans of his penis is applied to the inmost part of the woman. Thus the union is completed on both hands with the supremest degree of inducement. Furthermore, the moisture which flows round the parts, by its acrimony and saltiness calls out the whole papillary sense in the minutest details; for whilst the male member in general produces this effect by soft contact, this moisture produces the same by the individual touching of every single papilla; and thus multiplies the feeling. In the man however there is scarcely any sensibility of the kind, since the glans meets the urethra; of course therefore the female is excited by a more vehement passion than the man. From these considerations we may conclude, that the more remote or internal causes follow a like degree; for internals and externals always mutually correspond. The point is also proved by the fact, that the nerves of the female sex are more tender and lax, and thus more prone to assume that state by the first cause that influences them; the state being this,—that at the first signal given, or the first allurement that influences, the fibres expand more readily, and more readily vibrate; so also, in this ratio, the papillæ are more sensitive; and by consequence they excel both in quantity and quality in the female sex. The subjects of this passion themselves say and attest as much from fact and experience, and furnish us with confirmation: but as to the cause of this surpassingness of desire, we shall discuss it further on. Moreover as there is a series of causes that must concur simultaneously to the one effect; causes, that is to say, from the soul, from the mind, from the disposition, and lastly from sense itself, or the body (see this explained above); so there must be as many incitative or impulsive causes as there are causes in one series; for the first cause cannot produce the second immediately, but the superior merely communicates a nature to the posterior, whereby the latter, excited by its causes, rushes with greater proneness into the effect. The universal cause influences or flows into the posterior cause in no other way than by natural disposition or by inclination. Thus if there be a number of degrees in one series, there ought also to be so many impulsive causes (so the mediate causes are termed), that enter by the doors of the senses, and by the way of the imagination; and of ultimate causes also that come by touch, particularly of the members of the organ dedicated to transmission and reception; whence those delicious ecstacies which the whole world so vehemently loves.

141. *Sexual love, therefore, springs from the same causes indeed in both sexes; but nevertheless in its principles it is not exactly of the same nature in each. In the man, the first cause of this love is the propagation of offspring for the sake of society on earth, and more remotely for the sake of society in heaven; in the woman, however, though it is for the same causes, yet it is for these, not as in the principal cause, but as in the instrumental cause; to wit, in order that the woman may receive that cause not as native in her, but as to be implanted in her by her husband; and thus may perfect that which is begun by, and flows immediately from, her husband. These ends then in the two sexes correspond to each other, just as active forces correspond to passive, which, when they are united, produce one cause. Of this nature also is the love of the one towards the other, which love requires to be stronger in the female sex, because it is not excited by active but by passive principles. And in order to its suitable excitation, and that the woman may be subject to the active force, and to the will of the husband, it is necessary that her nature be such, that her will be subject to the affections of her disposition; but the affections to the will, and the will to the reason, of her partner. To bring this result about, excitements more numerous, and of more exquisite sense, are bestowed on women; which leave them*

less at their own disposal, but dependent on the right of another as their governor. From these facts we may conclude, to the laws of the conjugal compact as it has been instituted by nature; and we may see that it implies, that the wife shall obey, and the husband command, so that when they are of one mind, she shall suffer while he shall act. Otherwise order is perverted, and nature changed, and the husband naturally pardons his wife if she falls into faults arising from her mutability of disposition, too prone to motions and affections. Another consequence of this is, the more ardent love and constant affection of the mother towards her offspring; for she is excited more readily by internal and external causes; and it is native in her to conceive offspring, to form them, to bring them forth; for whatever is formal is hers, and whatever is essential is her husband's. Now the formal constantly excites, but not so the essential, which goes deeper, and ultimately perishes: and thus it is that even with the husband the formal has at last to continue the parental passion or *storge*.

142. *Moreover the external female organs are formed with the view of excluding the fœtus when conceived and matured; that is to say, for the sake of birth; the vagina being capable of expanding as the labor requires, and the other cavities of yielding and making a passage.* This is the reason why the vagina uteri on its inner surface is provided with such numerous semicircular folds, and with fissures, and on its external surface with cellular tissue: nay, this also is a reason for the carunculæ, nymphæ, and the fossa navicularis or scapha; for all these foldings give a faculty of expansion. At the time of delivery all the folds open out, and together go to constitute one grand canal: they all yield as soon as the membranes break, and the child begins to come down. To say nothing of the other appended parts, as the pelvis; or of the yieldingness of the bladder and the intestinal tube, between which two the vagina lies in its course. But the several parts tend again to their smaller calibre, and indeed to their pristine state of contraction, and in fact relapse afterwards by means of the aforesaid foldings: this we may conclude from the various circumstances of the case. *To say nothing of other natural discharges to which the same cavity affords a passage, e. g., the*

lochia and the urine. But I do not venture to treat further in detail of the several uses and the several parts, for fear I should perhaps hurt that modesty and chastity which ought to be observed: I now therefore pass to the uterus, in preference to continuing the present subject; for the inner parts assail the senses less than the outer.

CHAPTER II.

THE UTERUS.

143. Heister. "*The vagina* is a large canal, not greatly unlike, though more robust than, the rectum. This canal extends from the orifice of the vagina, between the bladder and the rectum, to which it is adherent, to the uterus. Its length, when it is not stretched, is about six or seven finger-breadths. Its calibre is rather larger than that of the small intestine; in certain cases, particularly at the time of delivery, it is apt to undergo great dilatation. Its orifice is narrower than the other parts, and contracted by a sphincter. It is of a membranous substance, and on the inside wrinkled, nervous, and papillary (so Ruysch has observed); and hence it is very sensile: on the outside it is muscular, and by this means is enabled to embrace the penis blandly in coition. It is connected in front with the bladder; behind, with the rectum; and above, it is attached to the womb. The wrinkles or rugæ of the vagina are not circular, but more resemble those in the jejunum. They are largest in maids, particularly in the anterior part of the vagina; in women who frequently transact venery they are of smaller size, and as it were rubbed down; and in those who have borne several children they are almost entirely obliterated. Their use is twofold. On the one hand they augment venereal pleasure both for the male and the female. On the other hand they increase the facility with which dilatation takes place in childbirth.

144. As there are lacunæ and oscula about the urethra, so also they are found here and there about the orifice of the vagina. They are often large enough to admit a bristle. The greater part of them come from the layer of glands called Bartholin's

prostates, which secrete a mucous liquid for lubricating the vagina, and perhaps for exciting the stimulus of venery. The sphincter muscle or constrictor of the vagina, is a series of muscular fibres, arising from the sphincter ani, embracing the orifice of the vagina, and inserted under the clitoris and its crura. In the same place a cavernous substance surrounds this orifice; and during venery, when this substance is distended with blood, the opening closes in, so as to affect the male penis with greater pleasure, and at the same time to be pleasurably affected on its part by the penis."

145. "The uterus or matrix . . . is a part situated between the bladder and the rectum. It is hollow, in shape like a pear, but a little compressed; and it is intended as the place for the generation of the fœtus. With respect to its connexions, its posterior part is free; its anterior portion is united with the vagina; and its lateral parts are attached by ligaments, which are twofold. The *broad* ligaments are membranous, and continuous with the peritonæum, and on each side they join the uterus and vagina to the walls of the pelvis. They consist of two membranes, which have a cellular membrane between them (as is easily demonstrated by inflating the parts) almost as is the case in the mesentery. The *round* ligaments arise from the upper part of the uterus, pass through the abdominal rings, and afterwards terminate in the fat about the groins. Each round ligament consists of two membranes, of a vascular plexus, of strong fibres which are especially conspicuous in and before childbirth. During coitus these ligaments are capable of drawing the uterus to the penis. In impregnated adults the uterus is about three inches long; in the upper part about two inches broad; one inch broad, below; and half an inch thick. In virgins it is still less than this; but in pregnant women it varies according to the period of their pregnancy. The upper and broader part is called the fundus; the lower part is the cervix; the remainder is termed the body of the uterus. In the latter, the vagina having been previously opened to expose it, we find presented for examination the os internum uteri or inner orifice of the womb, which stands out into the vagina almost like the glans penis in shape. It is very small in virgins; but in women who have borne children and in pregnant females it is a little larger, and

closed moreover in the latter cases by a very glutinous humor. In childbirth however, by a miracle of nature, it gives passage to a large fœtus complete. It is of a muscular substance, made up of various meshes of fleshy fibres, interwoven with vast multitudes of bloodvessels. In women who are not pregnant it is compact and firm; in pregnant females on the other hand it is spongy and full of sinuses, and susceptible of wonderful dilatation without derogation from its thickness. *(Comp. Anat.*, n. 236.) I have never found the uterus thinner, but often thicker, during the enlargement of pregnancy. *(Ibid.*, Not. 31*.) Externally it is environed by a strong membrane from the peritonæum; internally in the cavity (which is small in virgins), it is covered by a porous, nervous membrane, which seems to disappear in childbearing women. *(Ibid.*, n. 236.) Ruysch saw and has delineated in the fundus of the uterus of pregnant women, certain spiral or orbicular fleshy fibres, which he has called the new orbicular muscle of the uterus. To this muscle he ascribes the power of separating and expelling the placenta from the uterus after the birth of the child. . . . I have made search after this orbicular muscle, as occasion has presented itself, in the bodies of pregnant women; but I have not been able to find it as Ruysch alleges, but have detected only some muscular fibres stretched round about in various directions in an inextricable mesh. *(Ibid.*, Not. 31.*) The bloodvessels of the womb are tortuous, and form innumerable anastomoses. During pregnancy they are dilated into a kind of sinuses. They open upon the uterus and vagina by little orifices, and are the springs of the menstrual discharge. *(Ibid.*, n. 236.) Physicians are not agreed as to the exact spot whence the menses come, whether from the uterus itself, or only from the vagina. . . . Morgagni says that according to his observations, he has seen this discharge most frequently issue from the fundus of the uterus, but never from the vagina. . . . Littre confirms the position, that the menses arise from the uterus; so also Grassius, &c. *(Ibid.*, not. 32.*) The arteries of the womb are from three sources. 1. The spermatic arteries from the aorta: 2. The largest vessels are from the hypogastrics: and 3. There are vessels from the external hæmorrhoidal arteries; all of which three sets wonderfully communicate with each other, so that if air, wax, or mercury, be injected into any one of them, all the rest, even on the other side, are readily filled. The veins also are threefold, and have the same names as the arteries. They are destitute of valves, and are much more capacious than the arteries, particularly during pregnancy. Air driven in through them passes into the cavity of the uterus and vagina; and on the other hand inflation may often be effected of the veins from the cavity. The uterine nerves are from the intercostals and the sacral nerves. . . . There are little orifices with valve-like rugæ interspersed in the cervix of the uterus: these appear to be ducts secreting a glutinous liquid. Vesicles or globular corpuscules are sometimes observed in the cervix and orifice of the uterus. These contain a mucous humor. Many anatomists have regarded them as hydatids; some however consider them as the glands that secrete that glutinous liquid that closes the os uteri within during pregnancy. Others again reckon them as the new and true ovary in which the fœtus is made. Nor have there been wanting those who have called them the female vesiculæ seminales, and have supposed that a prolific semen is poured forth from them during coition. Their use is still dubious. They are the most numerous in childbearing and pregnant women; in whom also they are more conspicuous than in others. (*Ibid.*, n. 236.)

146. WINSLOW. "The uterus lies between the bladder and the rectum. . . . Except during pregnancy, . . . it is in adults about three finger breadths in length, one in thickness, two in breadth at one end, and scarcely one at the other. Its size varies with the age of the subject. . . . It is situated obliquely, the fundus being turned backward and upward, and the neck forward and downward; the broad parts lie next the rectum and bladder, and the narrow parts are at the sides. . . . In the natural state the internal orifice of the uterus is narrower than the duct of the cervix. . . . At the edge of this orifice, there are several small holes, answering to the same number of glandular granules, and which discharge a glairy lymph. The inner surface of the uterus is lined by a very fine membrane, which at the fundus or broad portion is smooth and even, but in the narrow portion which leads to the orifice it is wrinkled in a particular manner. The portion of this membrane which covers the fundus is perforated by a great number of considerable orifices, through

which small drops of blood may be observed to pass when the whole uterus is compressed. Sometimes it appears to be furnished with very fine little hairs, as though it were a villous membrane. Both these villi and orifices are more or less tinged with blood in those women who die during the period of the menses. In the narrow part, which answers to the cervix, each side is divided into two lateral parts by a kind of prominent longitudinal line, which is larger on the upper or anterior side than on the lower or posterior. On each side of these two longitudinal lines, there are certain obliquely transverse lines or rugæ, more or less irregular, and disposed like branches of which the longitudinal lines represent the trunks. Between and around these rugæ there are small lacunæ, through which a mucilaginous fluid oozes that closes the orifice of the uterus. We observe likewise in the interstices between the rugæ several transparent globular corpuscules. . . . The substance of the body of the uterus is spongy and compact, with a copious intertexture of vessels. Its thickness is nearly uniform on the sides and edges. The fundus is thicker toward the middle than toward the two angles. . . . The edges are likewise much thinner near these angles, but not much so near the extremity of the neck. The uterus is covered by a portion of the peritonæum, which serves it as a coat, and is the continuation of that which covers the bladder and rectum, running up from the lower and posterior part of the bladder, over the anterior part of the uterus, and from thence over the fundus, and down the back part, until at last it comes to the rectum. On the whole of each lateral part or edge of the uterus this portion of the peritonæum forms a broad duplicature, which extends on each side more or less directly to the neighboring lateral parts of the pelvis, forming a kind of membranous septum between the anterior and posterior halves of the cavity of the pelvis. This septum is afterwards continued in a loose manner to the peritonæum on the sides of the pelvis. These two broad duplicatures have the name of ligamenta lata, and alæ vespertilionum. The upper edge of each is partly double or folded, forming two small distinct duplicatures, which I term the pinions of the broad ligaments. . . . The laminæ of all these duplicatures are held together by cellular tissue, in the same manner as the other duplicatures of the peritonæum, and they contain the

Fallopian tubes, the ovaries, a part of the spermatic vessels, and of the vessels that go to the body of the uterus, the round ligaments also, the nerves, &c. (*Exp. Anat., Tr. du Bas Vent.*, 590—603.)

147. "The bloodvessels of these parts are of different kinds, viz. the hypogastric arteries and veins, the ramifications of which belong chiefly to the body of the uterus; the spermatic vessels so called, and the two vascular cords, called ligamenta rotunda, which might be more properly termed the vascular cords of the uterus, or of the broad ligament. The hypogastric branches, arterial and venous, arise from the artery and vein of the same name, gain the lateral edges of the uterus, and are distributed to all the parts thereof, both internal and external; they form a number of curves, and a multiplicity of interlacements. The arteries of one side communicate both upon the uterus and through its whole substance with those of the other side, and the arterial ramifications of each side form numerous anastomoses with each other. The veins communicate on each side in the same manner. All these bloodvessels communicate likewise with the spermatic vessels, with the vascular cords of the broad ligaments, and with the hæmorrhoidal vessels. These frequent anastomoses may be demonstrated by injecting or inflating the hypogastric vessels. . . . The extremities of these arteries terminate and open into the cavity of the uterus. There is this peculiarity in the veins, that they communicate with the internal hæmorrhoidal veins, and consequently with the vena portæ. The spermatic vessels have nearly the same origin in females as in males, and likewise the same course and interlacements. They never pass out of the abdomen, but are wholly distributed to the ovaria and tubes, and communicate with the hypogastrics, and with the vascular cords of the broad ligaments. The veins are very numerous in proportion to the arteries. They send out lateral ramifications which seem to communicate with the mesaraic veins and vena portæ. The vascular cords, commonly called the round ligaments, are two long small fascicles of arteries and veins interwoven and connected together by a fine cellular tissue; they glide into the great duplicature of the broad ligaments from each corner of the fundus uteri, as far as the annular openings of the abdominal muscles. Throughout this course each cord

thrusts outward or raises the anterior lamina of the duplicature, which consequently gives a kind of coat to these fasciculi, and makes them appear like distinct cords, connected to the anterior side of the duplicatures. They seem to arise from the communication between the spermatic and hypogastric vessels, and may be regarded as a particular continuation of the former vessels. . . . Afterwards they run in a course nearly resembling that of the spermatic vessels in males, pass out of the pelvis, through the openings of the abdominal muscles, and are lost in the fat of the upper and middle parts of the groins. It may be conjectured that these vessels furnish matter to the lacunæ, of which hereafter. As they pass out of the abdomen they are accompanied by a production of the cellular portion of the peritonæum, almost in the same manner as the spermatic cord in men, and they also bring with them a fasciculus of fleshy fibres resembling a kind of cremaster. . . . The nerves come from the lumbar, sacral and great sympathetic nerves, in the same manner as in males. The lymphatic vessels run chiefly in the coats continued from the peritonæum. I shall in another place explain the lacteal ducts, which are seen at an advanced period of pregnancy, and also the particular fibres which seem to be interwoven in the substance of the uterus in pregnant women, the innermost of which being disposed in a vortical manner, gave occasion to Ruysch to describe them particularly under the name of the orbicular muscle of the uterus." (*Ibid.*, n. 614—623.)

148. Boerhaave. "The woman has a broader os sacrum than the man; it also turns more out; and her os coccygis retreats further backwards. Moreover her ossa innominata are broader, with a greater distance between them, and in their lower part they turn far more outward. The same may be said with regard to the lower tuberosities of her ossa pubis, which likewise turn out. Hence women, unlike men, are the broadest about these bones; and the capacity of the pelvis is much greater than in males. . . . Women are also much flatter than men on the front part of the chest. . . . They commonly perspire much less than men.

149. "In the pelvis is situated the uterus; soft, pulpy, all vascular, readily extensile; not very elastic; almost free and unfixed: hardly subject to pressure, for it is defended by the

expanded membrane of the peritonæum. . . . It receives arteries from the spermatic and hypogastric vessels, or from the internal iliacs: the anastomoses twirl and curve into wonderful reticular plexuses, which surround the entire uterus, creep through its substance everywhere and on all sides, and at last vanish, constituting the chief part of the body of the organ. . . . The womb is pierced on its inner side . . . by a number of small orifices, from which there exudes a bland and aqueo-mucous humor." (*Inst. Med.*, n. 659—664.)

150. SCHURIG. "The vagina is contorted, somewhat oblique, and rather contracted; when not in coition its sides are somewhat flaccid, and collapse like those of the intestines; but on the other hand during the menstrual flow, as well as during coition, and in childbearing, the vagina becomes larger. (*Muliebria Historico-Medica,* 1729, III., 1., § 5, p. 200.) It comes up to meet a short penis, yields before a long one, dilates for a thick one, and contracts for a small one. [§ 7, 20.] *The uterus.* In virgins, unless the menses are on, the cervix uteri will only admit a fine stilet [§ III., 2. § 7, p. 218.] It is always closed particularly after coitus, and during pregnancy [§ 8 *ss.*, p. 219 *ss.*] Instances of enlarged uterus from various causes [§ 15, p. 232 *s.*] Double uterus, many examples of it [§ 31 *ss.* p. 260 *ss.*] A cartilaginous uterus [§ 35, p. 267 *s.*] A scirrhous ditto [§ 36]. Stony, callous, spongy ditto [§ 37, p. 270 *s.*] Cancerous [§ 38, p. 271 *s.*] Fungous, with fungus in the womb [§ 39, p. 272, *s.*] Moles, natta [§ 40, p. 275 *s.*] Wind from the uterus [§ 41, p. 276 *s.*] Flame from ditto? [§ 42, p. 281.] Hydatids in the uterus [§ 44 *s.*, p. 285 *ss.*] Calculi [§ 46, p. 290 *s.*] Hair [§ 49, p. 296 *s.*) Worms [§ 50.] Ulcers of the womb [§ 51, p. 299 *s.*] Gangrene [§ 52, p. 301 *s.*]

151. MORGAGNI. "In a girl of thirteen I found the posterior internal surface, not only of the cervix, but also of the fundus of the uterus, as it were in two parts, divided lengthwise by a deep furrow. (*Advers. Anat.*, IV., anim. xxv., p. 46. See also the remarks there cited from De Graaf.) 'In the last months of gestation," says De Graaf, "the uterus seems almost to put aside its membranous substance; for when dissected, it is so red in color, and is seen to be perforated by so many little foramina, (which are nothing else than the ends of divided

vessels,) that it really assumes the appearance of a spongy substance.' . . . Malpighi describes . . . that fleshy fibres massed together into fascicles, and forming a kind of reticulated structure, so make up the structure of the uterus outwardly, that its inner substance also is a congeries and structure of strong fibres; and even when the chorion and the placenta are pulled away, the uterus is found to be muscular underneath. This I saw very plainly, in company with some Venetian friends, in a woman who died two or three hours after childbirth in consequence of a mental emotion." (*Ibid.*, p. 47.) (Authors state variously that the uterus regains its former size, eight, ten, or sixteen days after birth. Riolan says that he has found this to occur in four and twenty hours; see Morgagni's account of this (*Ibid.*, p. 47.) In the body of the female mentioned above, although we found the external vessels of the uterus a finger-breadth in thickness, yet the sinuses communicating with these would evidently not admit the little finger in any part, and the foramina communicating with these again on the surface of the uterus, where a part of the placenta was still adherent, were such, that the largest of them by no device would give entrance to the tip of the little finger. It is therefore no wonder . . . that humor, or air, thrown in through the vessels, should reach the uterus in puerperal women. (*Ibid.*, p. 48.) We find it stated by nearly all anatomists, that the neck of the uterus is wrinkled and rough. I have not however seen it observed, that these rugæ, and this roughness are caused by valves, which are not turned upwards but downwards. . . . Sometimes fleshy and thick fibres, but more frequently several firm valves, produced from the doubled inner membrane of the cervix, are prolonged transversely or obliquely, between those fascicles of fibres, in such a way that they are firmly fixed by one border into the upper side, from which they spring, but the other side, which is free, is turned downwards. And hence it is, that when a probe is introduced, and pressed against them, they allow it to pass downwards with no trouble; but when driven upwards, they altogether prohibit its course the reverse way. . . . In the hedgehog the orifice of the uterus is girt up with a ring constructed in the same fashion; in the sheep also, and the cow, the cervix of the uterus is barred by a very similar structure. . . . The cervix of the uterus, or that tube which extends from the meeting of the cornua all the way to the vagina, rises into numerous and thickly placed folds which run straight upon it longitudinally. These folds are all raised more than in the rest of their extent, in one particular part, so that they altogether present the appearance of a ring. . . . The ring which is next to the cornua is narrower than that which succeeds it, and the rest are less contracted in proportion as they are more external. It is on this account, by reason of the barriers and valves facing downwards, that it is far more easy for a probe to get out of the uterus than to get into it." (*Ibid.*, p. 13.) Morgagni also says that he has found a great number of vesicles in the cervix, and sometimes also in the orifice, and in it alone (*Ibid.*, I., p. 44.) He also observes that on pricking the vesicles in one elderly female subject, no mucus came out, but a fluid like that which is contained in hydatids spirted into the air. (*Ibid.*, IV., p. 72.) "I have certainly always observed that the humor, or rather mucus, that I express in large quantities from those vesicles, was of a sluggish character; and when compared with the liquor of the hydatids that occurred occasionally in the vicinity of the testicles and the tubes, I found it to be quite dissimilar to it; whilst on the other hand I uniformly remarked that it had the greatest similarity with the mucus in the cervix, and in the uppermost part of the vagina. (*Ibid.*, I., n. 32, p. 44.)

152. Ruysch. "Description of a fleshy polypus in the uterus. (*Obs. Anat.*, Amst., 1721, VI., p. 9.) Of a schirrhous and malformed uterus, which hung out beyond the labia. (*Obs.* VII., p. 10.) Of the frequency of prolapsus. (*Obs.* IX., p. 11 *s.*) Also of inversion following childbirth. Prolapsus of the os uteri in childbirth. (X., p. 12.) I regard nothing as more certain than the fact, that the uterus performs great movement in childbirth: all the natural efforts of parturition are dependent to a great degree upon the uterus. In several puerperal women a few hours after birth, I have found that the motion of the uterus was so evident, that the nurses as well as the patients frequently used to ask whether there were not another child to come away. And on applying my hand over the belly, I found indeed the motion so remarkable that it seemed just as if my hands were thrust away by an actual fœtus: nay in some cases the womb rose up," &c. (*Obs.* XCIII., p. 96.)

153. Th. Bartholin. "The uterus is placed between the rectum, which lies below it, and the bladder which lies upon it. These are its two pillows. Why should we be proud, when we are born between dung and urine? . . . In women of the middling height, and who are in the habit of coition, it is generally about eleven finger-breadths long from the pudendum to the fundus: the fundus itself being about three finger-breadths. . . . In the last months of pregnancy, the uterus becomes two finger-breadths in thickness, and exchanges its membranous substance, for a red, fungoid and spongy substance, which is perforated with holes like a pumice-stone. . . . Its substance is thickest about the internal orifice, which is narrow. . . . The collum or cervix is attached by its own substance and by the membranes. The fundus is attached by peculiar ligaments. . . . So far as regards its own substance, the fundus is not attached, but is free. . . . At the sides it is connected by two pairs of ligaments. . . . The second and lower pair of these . . . pass to the fat and the membranes of the bones, and are obliterated near the clitoris, to which they are attached, and where they degenerate into a broad fine nervous film. (*Anatome*, cap. xxviii., p. 260—263, Lugd. Bat., 1673.) The collum or cervix is a narrow canal . . . which will admit a good-sized stilet, or a quill. . . . The cornua . . . which come forth from the sides of the fundus, lie one at each side, where the vasa deferentia are inserted. . . . The orifice, or internal os uteri, is oblong and transverse, but very narrow; but when it opens, it becomes orbicular and round. . . . Within the canal of the os uteri, there is an oblong tubercle added to its lower part, which closes up the orifice with greater exactitude: so says Riolan. He also remarks that around this tubercle we find pores or little orifices, which seem to be the extremities of vasa deferentia that terminate at the cervix." (*Ibid.*, cap. xxix., p. 376—280.) In *Ibid.*, tab. xlvi., fig. ii., Bartholin's shews "the lower and round ligaments of the uterus: . . . the vasa deferentia running from the cornua of the uterus to the testes; . . . also a duct arising from the vas deferens, and passing to the cervix uteri, whither pregnant women are said to ejaculate their seed."

154. Ant. van Linden [according to Schurig, *Muliebria*, p. 217] divides the uterus into os, cervix, and fundus (p. 218). Highmore [*Disquisit. Anat. Corp. Human.*, Hag. 1651, fol. 98]

observes that the os resembles the glans penis, only that it is a little blunter at its termination: it is perforated in the middle. PALFYN [*Description Anatom. des parties des Femmes,* Leide, 1708, p. 55 *s.*] says that in the middle of the collum or orifice is an orifice which is the beginning of a narrow duct that gives passage to what has to be expelled. It is always shut, except to give passage to the seed, to the menstrua, to abortions, or to foreign bodies. HENR. VAN SANDEN [*Observat. de prolapsu uteri inversi,* Regiom, 1722, § 2, p. 15] remarks that otherwise the cervix will only admit the entrance of a thin stilet. [Schurig, l. c., p. 219.] The os allows of wonderful expansion, like the gape of a puppy, or the mouth of a tench. [p. 223.] After coition the internal orifice of the uterus is accurately closed, as Gofey proved by examination with the finger, according to the relation of Hoffman: but this Schurig confutes. [p. 252.] Wesling shews that a particular duct runs from the testes and tubes to the ligamentum teres of the uterus, which descends on both sides from the sides of the fundus towards the lower parts, and perforating the peritonæum and the tendons of the oblique muscles, is connected upon the ossa pubis to the clitoris.

155. REGNER DE GRAAF. "Many anatomists with Harvey clearly shew semen in the cavity of the uterus after coition; that is to say, prolific semen: for several days I discovered no such thing; yet of course the semen must go somewhere. Fallopius avers that he has often found the most exquisite semen in the tubes. Frederick Ruysch discovered a white seminal liquor occasionally in the tubes of gravid women, or at least a substance which, he says, is similar to male semen in color and general appearance. Hoffman and Bohn, as well as Gofey, are in doubt upon this point. For De Graaf says that a humor different from semen sometimes besmears the tubes. Hence some have thought that the semen was conveyed by the vessels, by the pores of the membranes, and by the little burrowing orifices. They say that the most spirituous part, the seminal aura, the genital spirit, flies thither. Some say that there is a valve in the way which hinders the passage within the tubes, and prevents anything from being conveyed from the uterus to the oviducts." (*De Mulierum Organis,* vid. *Opera,* Lugd. Bat., 1677, p. 344.)

156. If *any point be still doubtful in anatomical physics, it is the mode in which conception takes place, and in which the semen is raised up into the ovaries. The general opinion has been, that a kind of genital or seminal aura, or vital spirit, in the first instance penetrates intimately either the blood of the uterus, or the membranes and their juices, and by this way flies into the ovaries or the small receptacles formed with a view to receive it, and thus impregnates the little ova. Others have held that little worms crept through the cervix into the uterus, and from the uterus into the Fallopian tubes, and by these finally gained the ovaries; and afterwards breaking the coats of the latter, wormed their way into some particular ovum. At the present time the common and settled opinion is, that the seed itself penetrates into the uterus, and therefrom through the Fallopian tubes into the ovaries. Some anatomists however believe that they have found a new ovarium in the cervix and orifice of the uterus. Each of these parties has many experimental proofs to shew, yet there is a preponderance of facts against it; to say nothing of improbabilities, which, where the authority of experience is absent, are decisively dissuasive. But let each hold to his case, and let its defenders only see whether it be in harmony with truth, and answerable to the laws of nature. But as it is our aim to conclude from the connexion of things, and from reasons based upon experience, we will slightly mention the opposing considerations, not however attempting by any invalidation of other opinions to pave the way for our own. Connexion of facts and naked truth are able enough, and will defend their own cause.*

157. *The passage of the seed through the uterus and Fallo-*

pian tubes to the ovaries, labors under the following difficulties: that the glans penis belongs no further or deeper than to the internal orifice of the uterus, where we at once meet with a narrow tube in the situation of the cervix, which is narrow in itself, and covered and as it were fenced with divers little rugæ and as it were valvulæ conniventes, which hinder the passage into the uterus, but not the passage from it. Respecting these valves, see Morgagni, who proves what we assert by many considerations: by the little tube itself, which in the state of contraction will scarcely allow a quill to pass: besides which, the disposition of the fibres and corrugations is such, that when what is called the internal orifice is extended and folded out by the glans penis, this tube is completely shut; though when any force urges upon it from within, whether arising from the menses, clots or fungous growths, or the fœtus, it opens. *On these accounts it is still a question with the learned who have had the best opportunities of experience, whether any seminal humor reaches the uterus itself, or not.* De Graaf says that many anatomists with Harvey clearly shew semen in the cavity of the womb after coitus; although after some days' search he has seen no prolific semen there: but as it is necessary, so he says, that it must go somewhere, he takes refuge at last in a passage of the seed through the Fallopian tubes. *And if it did reach the uterus, the point would still be undetermined; for there are in the uterus so many pores and orifices, small and large, and its surface is so evidently cribriform, that the uterus itself would in all directions absorb completely this humor which came into it, and the more fluid, spirituous and genital part thereof.* This is supported not only by the structure of the part, but also by the reason which grows out of that structure, viz., that only the purest part, (that, namely, that would burst through the little tunics covering the ovum,) exspires with the humor of the semen. *But the passage that leads into either Fallopian tube, at the place of its insertion into the uterus, is so narrow, that it does not seem possible that the seed should choose it in preference to another which is likewise open:* since it also*

* N.B. Our business here is rather one of interrogation than of refutation. We have to discover whether the seed enters the uterus, whether it enters the tubes, whether the cavity of the abdomen, whether one ovum distinctively: and thus always the question of *Whether?* is that which we have to put.

appears, that the same case exists here as in the tube of the cervix, viz., that the little orifice is completely closed as soon as ever the uterus swells up: nor has it yet been shewn so as to render doubt impossible, that any seminal humor has been found in it. The latter is indeed asserted by De Graaf and Ruysch, but they both admit that these tubes were also moistened with their own proper fluids, and that this humor was similar to the seminal. De Graaf says that a humor different from the semen sometimes besmears the tubes; and Ruysch asserts that a white seminal liquor is sometimes found in the tubes of pregnant females; or at any rate a substance similar in color and general appearance to the male semen. *Meantime the seminal substance which has passed into the tubes, would likewise pass through the lattice-work of the latter into their spongy portion, particularly that subtile essence sublimed from the seed, which were it ever to succeed in reaching the other and larger aperture, it would inevitably be diffused over the whole of the subjacent cavity of the abdomen; because however closely the fimbriæ were applied to the ovaries, it seems impossible that they could prevent it, when it got to this field, from making shipwreck in the cavity of the belly.* For the other extremity or fimbriated border of the tubes floats freely in the abdominal cavity, so that whatever passes through the tube is not determined into the ovaries except in the most general manner, in order that these, like the rest of the viscera of the abdomen, may be surrounded with fluid: nevertheless nature, in other respects, requires certain or determinate directions. *Furthermore, that seminal aura, or if the reader pleases, seminal humor, would burst through not only the general coats of the ovarium, or the peritonæum, which being cellular in those parts would disperse far and wide the essences of the seed through its cells; but it would also break through the proper tunicula of the ovum, and not of one alone, but of all; and thus would be carried in, not determinately, but promiscuously; and this also would fight against the order of nature, which supposes definite determinations.* Animal nature forms determinations everywhere, and thus builds the organism, nay, the simple fibres, for the very first essence (see above); thence builds compound fibres for the animal spirit; and vessels for the blood. That no certain determinations should exist for carrying the seed, or the spirit of the seed described above, would be contradictory to order, especially when the business of propagation was in hand. It is a well-known fact that a cellular production of the peritonæum surrounds the ovaries. *And besides, those who have taken a plain view of these contradictory circumstances, which would almost surpass the faith of those who believe in miracles, as they have inclined their minds to this side or that, have chosen rather to take refuge in the impregnation of the blood or nervous juice, or in conception as taking place in the uterus itself; leaving out of view the Fallopian tubes and the ovaries, as if they were mere appendages, and without any known cause for their existence and use: or again, enquirers have betaken themselves to the cervix of the uterus, averring that therein lies the new and true ovarium.* But we do not tarry over these matters: it is easy to refute, and to destroy: but to rear up better buildings, and to discover the truth, in this most intricate labyrinth, is a task indeed—"hoc opus, hic labor." But opinions are sufficiently refuted, and fall to the ground surely enough, by the bare setting forth of truer views.

158. *If we duly consider the several connexions of the parts, and the ducts coincident with them, it will be abundantly plain, that the grosser parts of the semen, those namely which are visible to the naked eye and to the microscope, never reach either the ovaries, or the ovum which is there, nor commence the work of conception, but that this destination and use is confined to its interior essence, pure, genital, spirituous and vital. For this therefore there ought to exist definite determinations, leading directly into some one of the ova, but not into more than one at once. And such as is the primitive or spirituous seed, such of course ought to be the fibrous determinations, which do not present themselves to the naked eye, and scarcely to the eye armed by the microscope; unless where a number of the fibres are collected into a group, and take the form of fasciculi.* This is the reason why we must proceed with the greater distinctness, and examine the several parts separately and in combination, and where experience shews any beam of light to guide us, there tarry, and out of that light, form our inductions and judgments, from the connexion of causes, by comparison with the methods in use in other parts of the body, and likewise from the correspondence between the organs of the female and the male sexes. *To exhibit therefore a*

general idea of the movement or advance of the genital seed, the process, which we are here in brief to expound, appears to be as follows. The seminal humor, driven forth into its first receptacle in the internal orifice of the uterus, is taken up by it, and at the same time by the tube of the cervix; after being taken up, it is shut in; and then, after some time, is conveyed away through certain ducts into the ligaments of the ovary, which the old anatomists called vasa deferentia; and thence by an uninterrupted path of little canals it goes to the corpora lutea, twisted in intestinal convolutions; and from these, with unerring determination, penetrates not into the ova, but into the one ovum. When the root of the seed is struck in this, the business of formation begins. Then the ovum, expanding to the due size, the common coat of the ovary is burst through in that place; and the ovum slips into the tubes applied to the ovary, and at last is rolled down into the uterus, to which it is attached by means of the placenta, and thus advances into the form of its own species. In this way the several genital organs of the female sex have their own peculiar offices allotted to them. The vagina is for the reception of the penis. The internal orifice of the uterus is for the reception of the glans penis. The same orifice, and also the narrow passage of the cervix uteri, are for the reception of the semen, and for its resolution into its purer and essential parts and globules. The ligaments of the ovaries are for the determination of the semen; and the corpora lutea together with the ova, for the reception and insinuation of the same: the Fallopian tubes are for the taking up and transmission of the little ovum; and lastly, the uterus is for the formation of the embryo. This appears to be the circle of operations. But to determine whether such is the case or not, we must institute a special survey and consideration of all the parts. Before doing this however, we will corroborate our statements in a general manner by the correspondency of the organs, and of their coöperation, which is passive and reactive, the coöperation, that is to say, of the woman's parts with the man's: and by this means we shall in a measure support the positions already laid down.

159. *The particular use of the genital organs of the female sex, besides being gathered from observations, and by induction from the connexion of parts, is especially manifest from the*

correspondency with those organs that exist in the male sex, and of which we have treated already. Thus it is evident that THE VAGINA *completely corresponds to the male penis, and this so well, that it quite accommodates itself to the dimensions of the penis; and in fact whilst copulation lasts, is a kind of outermost coat to the penis, and closely conjoined with it by a humor that goes between the two: add to which that it serves for exciting the delights of copulation as an excitative cause. Then further it is evident that* THE INTERNAL ORIFICE OF THE UTERUS *exactly corresponds to the glans penis, and quite accommodates itself thereto; this very orifice being the extreme boundary of the copulation. Thus the one enacts the part of the active force; the other, of the passive; and this inner orifice in other respects plays almost the like part as the vagina.* THE FIRST PART OF THE CERVIX UTERI, *which is narrow, and as it were tubular, corresponds to the male urethra, in which the male semen is forwarded; this being the border where the seed passes away from the man's power into the wife's.* BUT THE TUBE OF THE CERVIX UTERI, *which immediately succeeds the orifice, corresponds not only to the prostates, but to the vesiculæ seminales in the man; for this tube mingles in with the injected seed a particular humor squeezed out from its little oscula and vesicles; and what the humor of the prostate has united, this humor again unties or resolves and dissipates; enabling the seminal humor to exist in its nudity, such as it was in the vesiculæ seminales. The things that in the man have been conjoined, in the woman are again to be resolved into their principles, that the genuine spirituous seed may at length pass into the chamber of the [female] testes. A similar order is always observed in both; as there is conjunction, so there is separation, and resolution back to the first essence. There is* A SMALL VAS DEFERENS, OR DUCT, *not observed by all, which passes within the common coat from the tube of the cervix immediately to the ligaments of the ovaries. This vessel or duct corresponds to the vas deferens in the male; it leads the purified serum of the seed all the way to the first part of the ovaries by distinct little channels. Where this duct terminates in the part of the ovary next the uterus, there a kind of second vesiculæ seminales are met with; and at this place the semen appears to be still further purged, and reduced to its first and genuine essence, such as it was in the epididymides. Thence it continues its onward course toward the ovaries; and for this reason the ancients called this ligament the vas deferens. Thus this duct from the cervix to this ligament, and thence towards the ovary, corresponds to the vas deferens in the male. There are moreover under the ova* THE CORPORA LUTEA, CONTORTED *like little intestines, and which answer to the epididymides or parastatæ in the male; their shape and twisted appearance also are not unlike the latter; and the spermatic vessels carried to the ovary and Fallopian tube, supply not only the ova, but also particularly these corpora lutea. A similar essence is received by them, that is to say, the sperm reduced to its purer globules; consequently so prepared that therefrom it can easily get at the ova. Thus they are distant to the farthest possible extent from the other organs, and are ends; for the beginnings are in the ova; and in this manner the two in every way mutually correspond to each other.* THEN COME THE OVA, *whose appendices are the corpora lutea, which therefore correspond to the testicles, whereby what is elaborated in the testicles, is deposited in the ova as its receptacles: so that the testicles at one end correspond to the ova at the other, as active force corresponds to passive, or effective force to recipient force. Up to this point the correspondence continues, between the organs of the female sex and the male organs; but here the correspondence ceases, and with it the connexion also, and the immediate communication. The parts which now come on, as the Fallopian tubes and uterus, are organs and viscera proper to females; these now take the matter up, and continue the web of conception until formation is attained. Now then first, as a viscus proper to the female, we come in the succession to the Fallopian tube, which takes up in the embrace of its fibres in its own little mouth or osculum, the ovulum torn out from the female testes, and by a writhing circumflexion rolls it quite down into the uterus. Lastly comes* THE UTERUS *itself, which is entirely devoted to receiving not the seed but the ovulum, and to growing and forming the embryo. Hence it appears that the several genital organs of women have their own proper offices, and that these correspond to those which exist in men; and thus the animal seed from the beginnings of its course to the end, performs an entire circle, which may be called the circle of conception and generation. But to render the truth clear, we must needs treat of the parts severally,* viz., of the vagina, the internal orifice of the uterus, the cervix, the ligaments of the ovaries, and the round ligaments of the uterus, the ovaries themselves, then the Fallopian tubes; and lastly, of the uterus itself.

160. *The first part that we meet with is the vagina, which is the vanward path and the boundary extended towards the womb. Its proper office lies in taking up the virile member, and conveying it towards the inner orifice. Its second office consists in pouring on a glutinous humor, as the manifestation of an effort to a real union. Its third office is, to excite pleasure and passion for venery, as well in the man as in his spouse.* These three offices are concurrent, being no more separate than substances are separate from accidents, forms from qualities, and essential from impulsive causes. The first two offices produce union, while the latter excite or stimulate it; the efficient cause being therefore reciprocal in both sexes. *The vagina uteri applies itself completely to the male organ, as coat to body. It is so contrived, as to allow itself to be expanded and constricted, and to grasp the member tight: nay it applies itself not only to the breadth, but also to the length of the penis.* Internally it is beset with wrinkles, like the valvulæ conniventes in the jejunum, whose law it is that they are obliterated when the jejunum is expanded: see above: these folds arise from the tunics, and are not quite circular but interrupted, the longitudinal folds serving as a basis to the others: they are therefore naturally obliterated when the vagina is expanded. In virgins, and subjects where the vagina is comparatively narrow, these wrinkles are of larger size. *This disposition of the vagina appears to have the effect, that the narrower it is, the longer it is; but on the other hand, the ampler it is, the shorter it is: the arrangement of the folds, and observations also, prove that such is the case. Furthermore, in the venereal excitement, the vagina makes an effort to expand, and aims to come up nearer to, and almost to come to meet, the male organ.* In virgins, or those who have not known the other sex, the corrugation is more considerable, and the length of the vagina is also greater, but the calibre or amplitude less: in other subjects the contrary is the case. That the vagina is accommodated for receiving the penis, is evident

not only from its extension, as already noticed, and its amplitude, but also from its dimensions, in that it is six or seven inches in extent. *In coitu therefore it constitutes as it were the outermost coat of the penis, just as though it aspired to be of one body with the latter; for it is of that condition that it gives itself entirely to appliance.* If the penis were environed by still another coat than those which it has, this could be no other than one like the vagina. Its name of vagina or sheath comes from this fact. Not only its folds contribute to its application, but the rest of its substance also, and this substance is robust. Moreover, at its entrance or beginning there is a spongy tissue internally, and a cavernous externally; the entrance being softer than the passage, with a view that the approach may be comparatively easy, but that the other part of the operation may cost some little force. *Moreover, a glutinous humor is expressed, which serves to stop up the interstices of the rugæ, and essays to produce a kind of real unition; but it is provided that the two bodies shall only be one so long as the act lasts.* See above. It is well known that a viscid and glutinous humor oozes from the vagina; that is to say, that thickset crowds of pores and little orifices perforate the internal coat; as is plainly proved by squeezing between the fingers a section of the vagina: see also the observations on the subject by Morgagni, Heister, Winslow, and other practical anatomists. Now as this peculiar glue oozes from these ducts and orifices; and moreover the internal orifice of the uterus supplies a fresh quantity from its own proper ducts; and again as there is the humor, not seminal, but expressed from the male urethra, and which accompanies the seed on its outer surface, which part is returned by reflux from the internal orifice; and yet interstices remain between the rugæ, and at the same time on all sides where there is any fissure to be seen: as all this is the case, it follows, that not only are the several spaces filled in, but also the humor endeavors to as it were glue the one to the other; which however cannot be done, because the act is short, and combined with reciprocal and local motion. But meantime the gluey powers of the humor are proved by its viscid essence. *Thus it is wisely and well provided that at the time of coition nothing shall get up to the orifice and channel of the cervix, except the male semen: and most of all that no breath of air shall penetrate; the air being hostile to generation, and to the operation of the seed.* As will be shewn below. Nothing is more hostile to the womb than the air, which disturbs the whole work of conception and formation. In fact, the air is the force of the activity of nature in the last or ultimate sphere; with hardly any laws but those of the most general sort: whence animal nature would try in vain to weave any kind of organic web unless the air were first shut out. The ether however is admitted, for it is of a more perfect nature. *And moreover by this contact and filling up of parts, the vagina instigates not only the male organ, but also the subjacent spouse herself, to ardent love.* The vagina is not merely nervous, but papillary also. The nervous coat, which is the innermost, makes rugæ or wrinkles: consequently, when touched, it produces a kind of general sense or feeling; the papillæ however give birth to a more distinct and individual feeling. Thus when the male organ is set in motion to and fro, and the humor that is put in shakes the several fibrils or papillæ, a secret titillation must of course spring up, to equal the [general] sum, which may be reduced to calculation. The papillæ are sensitive in proportion as they are naked, and the more the rugæ are unfolded. *The glans penis in men is the only part which is sensitive in the way in which the whole of the vagina is sensitive in women. The more then the male glans is bathed about with bland enticements and touches; that is to say, the more of such enticements are put in its way, the more pleasantly it is affected: now such touches occur at once in the outer orifice of the vagina, and at the same time throughout the progress to the vagina in the undeflowered virgin: but afterwards only in the inner orifice, if the penis gets so far. Add to this, that the frænum, by drawing back the prepuce, raises this goad to the summit; the retraction exciting the whole of the tunics, tendinous, aponeurotic and nervous, to contract the penis still more strongly; while the fancy and imagination of the mutual venery come now into play, and excite the mind itself to the utmost degree of ardor.* The glans, and particularly its corona or crown, where the prepuce terminates, is papillary, and very sensitive: see above. When therefore like a blunt wedge it enters the vagina, the above part of it is affected to the last degree by contact with the ready folds and soft unevennesses therein. The prepuce also is papillary in a degree, and when laid back it too is sensible of the same delights; though these delights are apperceived in a general sense, but not in their distinct individuality, because they probably penetrate to the cerebellum. All this feeling is concentrated upon the frænum, turned back as it is under the urethra; and the frænum thus communicates it to the inner surface of the urethra, and at the same time to the outer coats of the corpora cavernosa or penis, and is compelled to a peculiar constriction. By this means the muscular and aponeurotic coats react more forcibly against the active powers within (see above), and more juice is expressed from the male urethra. *The greater the reaction, the greater the rousing or stimulation. The tunic of the penis is reactive, and limitaneous relatively to the extension of the corpora cavernosa penis. Hence by the above means more of the venereal juice is extruded from the urethra, and also from the prostate glands and vesiculæ seminales. This is the reason why the virgin vagina naturally brings more delights than the ample vagina of women acquainted with copulation: and why the greatest delights of all are felt in the first flower, when the tunic presents its hindrance, and through its breach affords a passage into the strait vagina.* Besides the cause already stated, to wit, that the imagination of the mutual venery kindles the mind, the narrow measure of the vagina also, and the continuous touch of all the parts, contributes actually to this affection of the body. The hymen indeed is but an obstacle, and a kind of fracture of the first septum. Nevertheless, at this time, it is nothing but fantasy that kindles all this ardor: violent touch does nothing, but extinguishes ardor: when however hope is in the work, and at length the vagina is gained, then comes increase of the passion from a bodily cause also; and this augments and completes it. For this reason it is, that we have the most utter love for her that we have deflowered, and that we prefer the hymen before other delights. *With regard to the hymen, it is the septum which bounds off between the respective regions of the male and female organs. It is placed in front of the external orifice of the vagina, and at the same time constricts this first threshold: and so long as it remains entire, entrance to that sacred house is forbidden. This door is shut until the period of puberty, to enable the internal genital organs to ripen the meanwhile, and to prevent them from being infested by anything harmful, even by the external air: for the same is the case here as in the male sex, namely, that the genital organs are not perfect, or inaugurated into their uses, until the spirit and blood run over with superfluity, and no longer require to be spent upon the growth of the body: but when the superabundance comes, so that without loss to growth and increment expenditure can take place, the fluids are then naturally derived hither to be excreted: and this is the reason why at that period, and not earlier, the menstrual flux begins.* In these matters then there is a correspondence between women and men: the superfluity itself causes the urgent passion for venery; it is derived not only into the vagina and the glands there grouped and situated, but into the several vessels both of the ovary and uterus (of which subject we shall speak presently), so as to meet the superfluous parts of the male. The organs themselves grow and increase together with this material: the organs being nothing more than the instruments of the fluid agents; thus the one is fitted to the other. But we will now pass on to the rest of the subject.

161. *The internal orifice of the uterus comes on just where the vagina terminates, and in truth is so constructed as to be capable of expansion and apt constriction, and of adapting itself exactly to the glans penis: so much so, that it seems to be formed for the reception of the glans, while the vagina is for the reception of the remainder or body of the penis.* This is confirmed by the anatomists; for though in virgins the part is sufficiently contracted, yet when it is expanded, and when it has been expanded, it exactly corresponds to the glans penis: this therefore is the extreme terminus whither the latter penetrates the belly. This orifice is confounded by many with the tube of the cervix, and hence it is often denominated the neck of the cervix: at any rate a small narrow foramen opens from it to the cervix, or the tube of the cervix; and the glans penis cannot be pushed on any further: for it can by no means open the foramen that is continued to the cervix uteri. This orifice is sometimes likened to the gape of a dog, to the mouth of a tench, or to the intestinal cæcum. *This orifice not only accommodates itself to the intruded glans penis, but it is also bathed in the same kind of*

humor as the vagina, and moreover in the seminal humor of the man: so that all passage is precluded save for the male semen into the tube of the cervix. Observers have also detected vast numbers of glands in this situation, and a peculiar humor, as in the vagina: moreover the humor that oozes from the urethra, and occupies the circumference of the seminal torrent, remains here in part, and sheaths the penis, and is carried out on its departure. *This orifice also excites passion both in the male and in the female, for they are so fitted together, that the universal papillary congeries on the male gland is in a state of contact, and moreover the frænum is quite drawn back, and the prepuce lies in the vagina.* These positions are similar to those that we lately brought forward respecting the vagina; for a like case is here met with; except that this orifice more strictly encompasses the glans, and touches all points of it.

162. *With regard to the cervix uteri, it comes on immediately after the orifice, and it is its first boundary: a kind of narrow duct, which admits no part of the glans penis; for it has a foramen opening from the before-mentioned internal orifice, which exactly applies itself to the orifice of the male urethra: in it therefore we find a continuation of the urethra, consequently a continuation of the seminal torrent, which pours immediately into this duct, but from which there is no open entrance into the uterus.* The reader will find this duct delineated, and its dimensions given, by De Graaf, Morgagni, and other anatomists. It is a kind of tube, so extremely narrow, that when contracted it will scarcely afford passage to a quill: see the anatomists. Moreover it is all full of folds, and indeed of valves, so disposed as to allow no passage inwards, but a free passage outwards, for example, for the menses, uterine masses, the child, &c.; see Morgagni and the rest. And then its fibres are so arranged, that when the orifice is forcibly drawn open without, this duct is more straitly closed, except the fine aperture leading from the orifice to which the urethra is applied. It cannot then be gainsayed, that it is a receptacle for the semen, and at the same time a shield or barrier to prevent the seminal humor from penetrating further. *Wherefore this first entrance to the uterus is the first receptacle of the seed, such as it is emitted from the vesiculæ seminales and prostate glands; at the same time it is also a*

bar to its passing in further, to the uterus. For to this point is the urethra continued; and to the urethra the cervix answers as a continuation of the duct; and into it the semen is ejaculated: all which is plain from the connexion of the parts. This is the very semen that comes from the vesiculæ seminales and prostate glands; wherefore to them this part corresponds. *Therefore this first part of the cervix represents the vesiculæ seminales in men; yet as it corresponds to them, and the semen is mustered in these vesiculæ, and again combined in the man by the humor of the prostates, and as it were coagulated, so in this place it has again to be resolved into its principles: in the first instance then it is here separated from the humor of the prostates.* And this, indeed, first of all by the throwing on of a comparatively thin or aqueo-serous humor (see Morgagni and Boerhaave), that resolves this thickish matter. Consult the anatomists respecting the glands and as it were hydatids that are met with in this situation: they all agree in affirming that numbers of these glands are discernible. This humor, which comes from the male prostates, appears to fall back into the subjacent orifice and vagina, the passage whither is patulous and headlong: nay sometimes it would seem to be absorbed in this situation by the veins. Occasionally a humor of the kind has been known to run out after the lapse of several days. *After this humor has passed away, there remains the true seminal humor, like what is wont to dwell in the vesiculæ seminales of men, and in fact it dwells here also among folds that represent gyres and convolutions such as there are in the male vesiculæ seminales.* It was on this account that some, particularly among the ancient anatomists, called this part of the cervix uteri, the vesicula seminalis of the woman, and chose to affirm that they saw a kind of new ovarium in this situation: nay this was the reason why they imagined, that the feminine semen here met the masculine, and to this therefore they also attributed a plastic force. Yet nothing more is found here than a liquid of but poor density, which resolves and dissipates from the seed those parts that have been thrown on it by the prostate glands. In order that the seed may be fruitful, it is necessary for it to return to its first globules or spirituous substances, and the return first begins here; for here the semen lies enclosed, it tarries, it is rolled about among the folds, and so by degrees it shakes off all that had been added to it.

163. *A peculiar filamentary and fibrous duct runs from this part of the cervix uteri. This duct terminates in the ligament of the ovary close under the insertion of the round ligament, or ligamentum teres, of the uterus. As this duct plainly opens under the general tunic of the peritonæum and the extension thereof all the way to the beginning of the ovary, this may afford ground for concluding, that the semen, thus purged from accessories, follows this path through peculiar ducts: and thence that this duct represents the first part of the vas deferens.* There are but few anatomists who have observed this duct; or probably the majority have not regarded it as of sufficient importance to merit a careful investigation and description at their hands. It has however been described by Riolan, and after him by Bartholin, who has delineated it in his plates. He mentions [p. 280] an oblong tubercle within the canal of the orifice, appended to its lower part, and which closes the foramen with great exactness. Around these pores and little orifices which are visible, he says that there are the extremities of vasa deferentia terminating at the cervix. In the description of his figure he says that there is a duct arising from the vas deferens, and which passes to the cervix uteri, into which pregnant women are said to ejaculate the semen. In common with the old anatomists, Bartholin means by the vasa deferentia the extremities of the ovaries that are inserted in the uterus: respecting which we shall speak presently. In order then that we may not break the thread of the present treatise, it will now be necessary for us to betake ourselves from the present heading to a description of the ovaries, and thus to complete the whole circle. The uterus comes last; for it is the end and as it were centre of all these operations. Afterwards, when we have come to the uterus, we shall again take up and weave in the thread which we now abandon, and discuss the use of the womb. In the meantime the memory must be carefully stored with what we have already said respecting the uterus, its ligaments, and spermatic vessels. It is an integral series, and unless I follow it straitly, it will be easy to lose the way.

CHAPTER III.

THE OVARIES.

164. HEISTER. "The ovaries, or as they used formerly to be called, the female testes, are two bodies, one on each side, annexed to the fundus of the uterus. They are in a measure globular, smooth, and whitish; and are connected, 1. With the fundus of the uterus, by means of the ligamentum teres, which the old writers called the vas deferens of the woman, but which has no passage leading to the uterus. 2. With the Fallopian tube and the sides of the pelvis, by the broad ligaments of the uterus and the alæ vespertilionum. And 3. With various other parts by means of the spermatic vessels. In shape they approach to oval, but gibbous above, flat below. Their size differs according to differences of age and temperament; they are very large (often nearly two drachms weight) in the vigor and prime of life, and in libidinous people, and frequently decorated with prominent vesicles: in old women however they are dried up, small, wrinkled, hardly half a drachm in weight, and disfigured by cicatrices or scars. The alæ vespertilionum are membranes which intervene between the ovaries and the Fallopian tubes; they have a cellular substance lying between their folds, and are encompassed by a strong, white membrane derived from the peritonæum. In point of substance the ovaries are membranous, fibrous, reticulated, with vast numbers of vessels woven into their texture. Among these vessels we often see round vesicles, more or fewer, and according to the differences of age or temperament filled with a humor similar to the white of egg: moreover this humor, by boiling, obtains a consistence, color and taste similar to that albumen. This analogy with eggs,

led Steno to call the vesicles ovula; the largest of them being hardly so big as a pea. Sometimes as many as ten, fifteen, or twenty, or sometimes even more ovula, are seen in one ovary; sometimes, on the other hand, we cannot find more than one or two. These ovula are thought to contain the first rudiments of the fœtus. Hydatids also are often found in the same situation. These however are morbid, and frequently are the causes of ovarian dropsy. The corpora lutea consist of a kind of glandular substance, of a yellowish color, and twisted like very delicate intestines: they lie under the ovula, which are set in their substance. Occasionally they are conspicuous in virgins, but chiefly in pregnant and child-bearing women." *(Comp. Anat.*, n. 237.)

165. WINSLOW. "The ovaries are two whitish, oval, flat, oblong bodies, situated at the sides of the fundus uteri, to which they are attached by a kind of short round ligament, and together with this ligament they are enclosed in the duplicature of the posterior pinion of the broad ligament. They are composed of a very compact, spongy substance, and of several little transparent balls or vesicles which are called ova. The spongy tissue surrounds each of these vesicles very closely, and seems likewise to furnish them with distinct spongy coverings or calices. These vesicles are to be carefully distinguished from other abnormal ones, termed hydatids. The ligaments of the ovaries lie in the edges of the posterior pinions of the ligamenta lata, much in the same manner as the umbilical vein lies in the edge of the anterior or umbilical ligament of the liver. They are round cords of a filamentary texture, attached by one extremity to the corner of the fundus uteri, a little above and behind the level of that fundus. They were formerly believed to be hollow, and were regarded as vasa deferentia." *(Exp. Anat., Tr. du Bas Vent.*, n. 604—606.)

166. BOERHAAVE. "The ovaries . . . are two scabrous bodies of considerable size, spheroidal in figure, with uneven surfaces, and each enclosed in a firm membrane. They are supplied with blood by the spermatic and hypogastric arteries as well as other vessels; all of which are so closely united before they enter these organs in this situation, that they seem to constitute a single vessel, as it were a woof of reticular plexuses. . . . The

same is true of the corresponding veins, as well as of the numerous lymphatic vessels. The vessels are so mingled with nerves to make up the structure of the ovary, and form such an intricate mesh, that the resulting fabric hardly admits of description. But on the surface of the ovaries, under the membrane investing them, we find certain spherules which are attached below by a thickish calyx to the substance of the ovary: these spherules are pellucid, and filled with a lymphatic fluid which coagulates when subjected to heat; and they consist of two concentric little membranes, but which are connected together in the closest manner. These ovaries are present in all healthy female bodies; in girls a few years before puberty, they are small; but they gradually increase, and at the period of prolification they are of very considerable size: in faded and withered subjects again the ovaries are observed to be diminished. The little bubbles or vesicles already mentioned, adhere by their calices to the exquisitely minute ends of the blood-vessels. After the exercise of venery they are just apparent; then they gradually swell, stage by stage become more pellucid, thicken in their membranes, raise the membrane of the ovarium, stretch it, and dilate it into a papillary form: by this means they stand out from the surface, hang from their peduncles, separate from them, leave a concave cicatrix in the torn substance and membrane of the ovary, but which cicatrix gradually tends to obliteration. It remains to be added, that in these bullulæ [or ova] while still adherent to the ovary, fœtuses have been observed." *(Inst. Med.*, n. 669.)

167. Morgagni. "De Graaf not only teaches that shortly after coition a peculiar glandular material enveloping the entire ovum, and of a globular shape, grows between the membranes of the ovum; but he also says . . . that ova ten times smaller than before coition, are expelled from this globule, that is to say, from the corpus luteum, into the tube." *(Advers. Anat.* IV., anim. xxviii., p. 51.)

168. Schurig. "Cases of large ovaries, some as big as a child's head, some the size of a fist; the enlargement due to ascites, and dropsy. *(Muliebria Historico-Medica,* 1729, III., 3, p. 314 *s.*) Cases of calculus of the ovary: of hair found in it: of suppuration of the ovary, and of absence of the part.

(§ 9 *ss.*, p. 316 *ss.*) Oviparous animals have one ovary, viviparous animals have two. In ovipara the ova or eggs are contained in no common membrane, but are heaped as it were in a bunch. (§ 17, p. 331.) The ova have a cicatricula: some one has said, that the whole part deserves the name of a cicatricula." (§ 18, p. 332.)

169. THE *genuine seed then, such as it exists when elaborated in the vesiculæ seminales of the man; cleared also from the humor of the prostates in the tube of the cervix; rises at once through a kind of vas deferens all the way to the ligament of the ovary, namely, to that cord which the old writers regarded as the vessel conveying the semen of women.* The reader may see this first part of the vas deferens running from the cervix uteri to the thin part of the ovary, delineated by Thomas Bartholin, in his *Anat.*, tab. xlvi., fig. 2, letters *k k;* and the insertion shewn at letter *e.* The vas deferens in the ovary, also in the same tab., but fig. 1, H H. *But as anatomists make little or no mention of the place of insertion of this first vas deferens into the other, we cannot do more than conjecture, that in this place also this masculine seed contrives a little delay, and purges itself of the remainder of that matter we have spoken of above; that is to say, of the body wherein the genuine spermatic globules are carried about, reduced to their form, defended from injury, and so prepared for use. One ground of this suspicion is, that the anastomosis is oblique in this place: where also there is a duplicature of the broad ligaments, and a coalition of that vas deferens with the round ligament, which so much consists of spongy substance, that the ancients believed these ligaments to be hollow. Whether in this place there be any space for the reception of this semen, and for further clearing it of its matter, is a point which we cannot decide in the affirmative, though probability may be conceded. Furthermore, the part secreted, and cleared away from the true testicular semen, can at once be sent away there through the round ligaments, and so be despatched to the region of the vulva, and at the same time to the fat of the pubes. The connexion of the parts*

argues thus much; likewise the vicinity of the round ligaments, their complication or folding together, their spongy nature, their defluxion to the groins, and evacuation there: perhaps also some lacunæ devoted to evacuation may be seated about the clitoris. Where experience is wanting, the connexion of things supplies material for judgment: meanwhile, time has to shew whether experience takes the same side. Certain it is, that these ligamenta teretia are spongy, are at last inserted in the os pubis under the clitoris, and thus can take up all those recrementitious portions, that the remains not only from the ovary, but also from the purified semen, throw off. The case here is similar to what happens in the vegetable kingdom, in that fruits of all kinds, after falling down into the ground which is their womb, in the first instance are freed by maceration from all adscititious materials, as in the present case by acts of severance or diremtion. Now as there are two vasa deferentia or deferent vessels, so also there are two spermatic vessels, one the continuation of the other. The one spermatic vessel which is applied to the testicles: the other is the round ligament, which like the spermatic vessel in men is continued through the muscles of the abdomen outside the uterus: so that here there also is analogy and correspondency between the two cases. *The sperm thus stripped or denudated, that is to say, separated into its primitive globules, is conveyed onward through the filamentary cord, which at the present day is termed the ligament of the ovaries, but which the ancients called the vas deferens of the woman. This ligament is composed of longitudinal threads, and it is a bundle of ducts, and is exactly suited for the determination and carriage of an essence of the above character.* This is evident from the mere description of these ligaments: see Winslow: nor do we need to ask here for any cavity, any more than in the least channelets of the fibres: they are then determinations both of a pure and subtle essence. *At length this seminal material, thus cleared down, is determined and carried away to the ova, or to its own little ovulum, and indeed to the corpora lutea, which lie like cups under the ovules or vesicles, and receive this pure semen, very much as the epididymides or parastatæ about the testicles of men, to which therefore they exactly correspond. They are almost similarly convoluted to the latter, and similarly cause this seed to circulate through their doors; as appears not only from the connexion of parts, but also from actual experience in pregnant women. For these corpora lutea swell up during the first days of pregnancy, and almost surround the enclosed ovulum, cherishing and embracing it on every side; thus supplying and forwarding to every point this spirituous stranger and genital guest.* On these subjects see De Graaf and the other anatomists, and consult also the descriptions of these corpora lutea. *Add to this, that the purest blood of the spermatic vessels arrives hither, just as to the epididymides in men.* The spermatic vessels together with the hypochondriac pour their blood around this whole mass, and by their capillary ends enter the corpora lutea: not to mention also the nerves or fibres, which form wonderful meshes in this situation. *Lastly, the ovula themselves correspond in women to the testicles in men, for they take up that same essence within their membranes, or within the membrane, from the corpora lutea or epididymides: they are therefore the ultimate receptacles, and are placed at one extreme, while the testicles occupy the other; and the two regard each other mutually as ends; or as active and passive forces regard each other; or as efficients and recipients; or as essence and form.* This indeed follows from the nexus of parts, and frequent experience plainly confirms, not to say proves it. Thus the ovula or corpora lutea have very often been observed to swell up, to be there swathed in their first coverings, and as it were to undergo incubation. The corpora lutea give the first nutriment; then the nourishing comes from the Fallopian tubes; and lastly from the uterus. Whoso denies this, denies experience. *Yet the ovula are only receptacles, and are scarcely anything more than a mere cicatricula, like what there is in bird's eggs: nor is more juice supplied to them than suffices for those first moments: and on this account the ovum undergoes no increase, until it is quite filled with the new spirituous guest, that is to say, impregnated.* In a short time it consumes these first-fruits of humor; for it emulges its nutrition, not from its albumen or vitellus, but extrinsically from the corpora lutea, the Fallopian tubes, and the womb, that is to say, from the parent. *In this way also we understand how not more than a single ovum is impregnated in the human species. For the vasa deferentia, all the way from their first junction with the ligaments beside the uterus, are filamentary, and every thread is a determination to its own vesicle. The thread then which is first opened, or which corresponds to the ripest ovum, must be opened of necessity. Whereby the rest of the threads, not yet sufficiently open or gaping, are shut and closed by the swelling of the first. Thus open way appears to be allowed to only one ovum. It would be otherwise if the seed approached from without to the ovaries, and impregnated the ova before the corpora lutea which are their base; which however is contrary to experience.*

CHAPTER IV.

THE FALLOPIAN TUBES.

170. HEISTER. "The Fallopian tubes . . . are two canals, one on each side, almost conical in shape, tortuous in direction, and connected to the fundus of the uterus. . . . They enjoy a close and continuous connexion with the uterus, and a loose connexion, by means of the alæ vespertilionum, with the ovaries. They are six, seven, and sometimes as much as eight or nine finger-breadths in length. In the middle they are of about the thickness of the little finger. Their extremities are more narrow, and indeed the ends by which they join the uterus are very small, and open into its cavity: these will give passage to air, or to a fine stilet. The uterus then, like the bladder, has three orifices to it. The other ends, floating freely in the abdomen, are more ample, and will admit a moderate sized tube; and they are decorated with a number of fimbriæ or fringes, as it were of a muscular character, and which when need is, apply themselves to the ovaries. . . . In substance the Fallopian tubes are membranous and cavernous, and they consist of two membranes. The outer of these seems to be continuous with the peritonæum, but the inner, with the internal membrane of the uterus. The latter is wrinkled, and covered with a lubricous humor; but it is not cellular, as in animals. The Fallopian tubes have vast numbers of vessels constituting a corpus cavernosum within their membranes; and by this they are capable of being erected and made stiff when they are to be applied to the ovary: the tubes also have moisture supplied them from these vessels. . . . In gullinaceous fowls and other birds these tubes have the name of oviducts." *(Comp. Anat.*, n. 238.)

171. Winslow. "The Fallopian tubes are two flaccid, conical and vermiform canals, situated more or less transversely on each side of the uterus, between the fundus and the lateral parts of the pelvis, and enclosed in the duplicature of the anterior pinions of the ligamenta lata. Each of them is fixed by its narrow extremity in the corner of the fundus uteri, into which it opens, but by so narrow a duct, as hardly to admit a middling silk thread. From thence their diameter augments by degrees all the way to the other extremity, where it equals about the third of an inch. The body of each tube runs a winding course, and their large extremity is bent toward the ovary. These large extremities are irregularly rounded, and terminate by a narrow orifice, a little plaited and turned toward the ovary, where it presently expands in the form of a membranous fringe, full of plaits and incisures. These fringes are called the pavilions of the Fallopian tubes. The fimbria is not of equal breadth throughout. Its circumference is in a manner oval, and the longest segment of the fringe reaches to, and is fixed in the ovary. The folds are disposed like laminæ on the concave side. The tubes are composed of fleshy fibres, some of which are longitudinal, and some obliquely circular, with an intertexture of another very fine substance. The anterior pinions of the broad ligament serve as a point of attachment, and as a common or external coat to both tubes, in the same manner as the mesentery does in relation to the intestines. This renders the tubes, and especially their fringes, more or less floating, and causes their direction to be very imperfectly given in most of the figures. Their cavity is lined by a soft and as it were glandular membrane, which is plaited longitudinally, almost like the inner surface of the trachea. These folds are stronger and broader near the great extremities, than anywhere else. They seem to be of a spongy substance, and the interstices between them are moistened more or less by a fluid, which is continually oozing forth." *(Exp. Anat., Tr. du Bas Vent.*, n. 607—613.)

172. Boerhaave. "From the two upper orifices of the uterus there proceed two membranous canals, of small calibre near the uterus, but which gradually dilate, and bending into cellular semicircles, end at length in a broad orifice whose margin is set about with fleshy fringes. They consist of an inner

glandular membrane with strong fleshy bands. During the period of the venereal orgasm in nearly all kinds of animals these tubes swell, stiffen, and blossom with carunculæ. In the bodies of women submitted to dissection the tube has often been seen embracing the ovary in its fringes. And moreover in the cellular recesses of the tubes fœtuses have been found in various situations, also at various periods after conception has taken place, and therefore varying in size." *(Inst. Med.*, n. 668.) And respecting the passage of the ovum from the ovary to the womb, Boerhaave says, that the phenomena of false conception and abortion, and the presence of fœtuses in the cavity of the abdomen, &c., &c., prove the view which he takes of it, as well as of the nature of the ova: see *Ibid.*, n. 671, 672.

173. MORGAGNI. "On a first glance and examination I have often thought that the tubes were closed in some part, and particularly at the uterus, whereas in reality, on careful examination and dissection, I have scarcely ever found them closed. *(Advers. Anat.* I., n. 30, p. 40.) After the tubes have mounted the dorsum of the female testes, . . . they frequently pass down behind the latter, and hang forward below them with their orifices looking sometimes downwards, sometimes upwards. *(Ibid.*, p. 41.) On careful enquiry instituted upon the bodies of twenty women, young and old, virgins, wives, and mothers, thirteen of their tubes (often those on each side) were found to be curved in such a manner, that they first mounted above the female testicles [the ovaries], and then descended behind them; the orifices of both of them being for the most part turned downwards: that of one however, and sometimes that of both, being occasionally pointed upwards. *(Ibid.* IV., anim. xli., p. 74.) Of all the animals which I have dissected up to the present time, the dog and the hedgehog are the only ones in which the terminal membranous expansion of the tube was fashioned into a sacculus or little bag, garnished on the inside, if my memory serves me, with fatty stripes, and loosely embracing the ovary. . . . Although Wharton, as De Graaf relates, found the tubes of a pregnant mare provided with great numbers of valves throughout their whole extent, yet I must say that in the body of a mare not in foal I have been unable to find one single valve in the tubes." *(Ibid.*, p. 54.)

174. Schurig. "The old writers gave the broad ligaments various names; they called them antennæ, spiramenta or breathing holes of the uterus, oviducts, cornua, also vasa deferentia." (*Muliebria Hist. Med.*, iii. 3, § 1, p. 365.) They are attached to the testes by nothing beyond a most delicate membranous connexion: De Graaf. (iii., 4, § 5, p. 370.) Cases of tubular conception. Ruysch found the tubes or oviducts sometimes distended, and deep red on account of conception. (§ 8, p. 375.) De Graaf mentions that the tubes have been found closed. Ruysch reckons the closure of the tubes as one of the causes of barrenness; and if they are closed at their extremities, he generally found them firmly adherent to the ovaries. (§ 9, p. 376.) Caspar Bartholin says that they are closed by valves in the rabbit, and that the passage cannot be forced by air blown in with ever so much force; nor by water injected by a syringe from the side of the uterus. (§ 10, p. 378.) It appears probable to Ruysch, that the menses and other heterogeneous humors may escape by the tubes into the cavity of the abdomen: he adds an observation of the fact: blood likewise he considers may pass by this direction. (§ 12, p. 380.)

175. The *Fallopian tubes receive the ova from the testicles a few days after coition, and convey them along their canal towards the uterus: as mere anatomical experience confirms and proves. For anatomists have seen the tumefied ova, also the external tunic ruptured, and the cicatricula or scar left behind: moreover the fimbriæ or fringes with their little mouths erected towards the ovarium, but the ova themselves at various distances from the uterus. Then again they have seen fœtuses in the tubes, and others which had escaped into the abdominal cavity, not to mention other phenomena; from all which it has resulted, that belief needs nothing unless actual sight, to empower us to state, that good observers have seen the passage into the tubes, the transmission through them, and the passage into the uterus.* All these circumstances, brought together, compared and confirmed by so many trustworthy authorities, leave no longer any room for doubting that the Fallopian tubes are oviducts: though with respect to the passage through them of the material seed, or even of its spirituous part, there is just right for hesitation. *This much also is allowed, that the aforesaid tubes, at the time when the ovum separates from the ovarium, apply themselves, and their fimbriæ, to the ovarium, and embrace it so closely, that the ovum can do no other than roll down into that orifice.* The erection of the tubes is a fact that has been seen and observed by many, and this, indeed, most generally about the third day after impregnation: moreover, when inflammation of the tube occurs, a similar kind of application exists: and we may infer from this, that the union at the time of which we are speaking, is very close; and of such a character, that the ovum is bound

to run into the orifice by these finger-like determinations of the fimbriæ. But there are many reasons why these tubes are distant from the ovary, and in their flaccid condition are rather turned downwards than upwards; also why they are not attached; and why they are erected, and conveniently adapted: and why the ova do not more often escape into the abdominal cavity. All these particulars are by no means without their causes; for nature, especially in the animal kingdom, does not construct the least hair's breadth excepting always for causes, and for ends: indeed there is not the least thread of vessel, or direction of fibre, that is not for the cause or use of the parts that follow it.

176. *As soon as the spirituous part of the sperm has arrived at its genital bed in the corpora lutea, and thereafter at the little doors of its ovum, it at once girds itself up to the work of holding and forming the thread of its new life and new body; and summons to partnership in the operation all the genuine spermatic portions through the threads of the vas deferens; and so it commences the web. It is the inherent nature of this essence, that wherever left to its own free action, it intimately conspires and endeavors to form this web; for it regards that which is to be formed, as formed, and the last ends in the middle and the first; and there is nothing but ministers to it, whilst it marches from the prime stages of principles to extremes and effects. When therefore it has gained this first position, it at once sets to work, and proceeds thenceforth just as it would proceed in a body already formed: for there lies in it a most perfect representation of the latter.* But on these subjects we have spoken above; also in Part I.; and still further in our Psychological Parts; and where we treat of the Soul.

177. *And as it resides with such a nature in this its centre or principle, it looks around it upon all things, whether near or distant, in the body of the mother, as its property; and from this property it takes out all that which will conduce to the formation of the new body. First therefore it demands from the fibres the most purified spirit, and from the vessels, the purest blood; for of these elements it constructs the tissue of its primitive form. Yea, whatever the extreme demands, all things in that microcosm must supply: thus if it demands anything from the fibres, or any*

thing from the vessels, the thing demanded, be it wheresoever it will, in whatever part of the nervous system, or of the sanguineous, is bound to contribute its share to the call. First then it demands from the spermatic vessels the purest essence of the blood; and also from the fibres of the sacral nerves, par vagum, intercostal nerve, in a word, from the cerebrum, spirit; and whatever this spirit brings with it: and all this it is empowered to elect and appropriate, so far as may be suitable. This is the reason why one or two days after impregnation, the whole of the system, both sanguineous, nervous and membranous, undergoes a remarkable alteration. Some persons might think, that so tiny a portion of blood and spirit as impregnates the ovum or its corpus luteum on the first day after coition, could not induce so great a change. But as only its most spirituous part, and its purest, or the intimate extract of the blood is required, that is to say, its quintessence, and thus its supreme quality, not quantity, it follows as a matter of course, that the whole mass of the blood must in a manner concur to supply this. Only from a great volume can a little of the most subtle essence be extracted. On the first days of all, nothing but the purest material is demanded. This must be supplied by the entire mass of the blood, and particularly by the spermatic vessels, which then for the first time begin to open their purest passages, and to operate. For this reason it is that the blood of the whole body communicates with the blood of the uterus, Fallopian tubes and ovaries, and the spermatic vessels with these parts immediately: for the blood of the spermatic vessels does not come from the mid aorta, but also from the iliacs or external hæmorrhoidal vessels, as well as from the hypochondriac, thus from the whole mass of the blood: likewise from the several fibres that are proper to the cerebellum, as for example, the par vagum, great intercostal, and sacral nerves. Now then this place where the ovum resides, and wherein there is a kind of new beginning, is to be considered as a peculiar centre; and whatever lies in the body of the mother is to be looked upon as circumferential, and as the periphery whence the centre can draw its legitimate and purest essences. This is the reason why the whole mass of the blood, and the whole spirit of the fibres, is now at work with the utmost energy, as it were boiling hot. Add to this that the soul itself, which regards effects at the same time with ends, and sees them simultaneously in one series as if present, and is distinctly conscious of this mutation, now bestows all its powers, and into this supply, pours a kind of new life: and therein sees itself as transferred and renewed, a second soul into which it transcribes its life, to hand it down to posterity. At this time then it represents to itself the already formed or material body as a kind of instrumental cause, which exists for the sake of this new progeny: thus it endeavors to bestow itself utterly upon the propagation of the latter. Here again we note a correspondency with a similar affection in men; who, under the venereal goad, make effort to pour forth all their spirits or life into the new life that is to be propagated, and thus to found a kind of colony as another likeness of themselves. The same thing is the case here; and now the soul strives to project all its life into this new soul: whence at this time the soul is busy to the last degree; although this does not strike our mental consciousness, except indeed in ultimate effects or facts, which are conspicuous enough. The whole work of generation is performed within nature, and without the consciousness of our rational minds. Thus the cause enables us to draw conclusions on the subject, not anatomically, but physiologically. *Whilst then the spirit and the blood in this considerable alteration of state, are in a kind of inmost fervor, and a slender quantity of the best quality is to be drawn out from the universal mass of the blood; the soul with all its forces engaged in the affair; it is needful that abundance of blood should stream thither from all directions; and thus inundate as it were, not only the ovaries, but also the Fallopian tubes, and the uterus as well; and this, in the ratio of the quality and quantity required. Hence it comes to pass, that the several parts are brought into a state of greater hardness and swelling, and those that were previously flaccid, are erected. In this way the Fallopian tubes, which otherwise hang forward freely in the abdomen, and incline downwards, are contracted; and nicely apply themselves, in the direction of their fibres, even their motive fibres, by their extremities to the place of the ovarium: and present themselves as instruments of receiving the little ovum that sails into them.* These positions are confirmed by the facts of the case. Thus the Fallopian tubes gradually erect themselves to embrace the ovarium: a phenomenon which has been observed by many anatomists: and this the tubes do, not at once, but by degrees, as the ovulum fills out, and prepares to undergo its first birth or travail: that is to say, as first the purer, then the grosser blood, runs copiously to them. The same thing is also confirmed in inflammation of the ovaries, in which, according to Ruysch's observation, the tubes are completely erected, and apply themselves to the ovaries. In the first instance the innermost parts especially are disposed to this mutation by the purest nervous fibres, which are inserted in great quantities, and by the membranes: in the next the parts are disposed to the same by the larger and grosser fibres and vessels; and at last, by the muscular fibres, of which also the tube is reckoned to consist. And furthermore when the blood fills up the cavernous substances, (as it does in the penis, of which we spoke above,) or the septa between them, then an erection of the whole follows; and erection also of the fleshy appendages or fimbriæ. For they are so twisted and as it were convoluted around the entire extent of the ovarium, like the turns of a spiral shell, that when contracted they cannot at once apply themselves to the ovarium; on account namely of the convolutions. Thus at any rate all things dispose them as a matter of course to the end foreordained by nature.*

178. *These tubes cannot fail to have a double motion: the one motion, pulsatile, according to the times of the heart's beating: the other motion expansive and constrictive, according to the times of the abdomen and peritonæum, that is to say, of the breathing of the lungs.* An abundance of blood creeps through them, and is everywhere propelled, to circulate always anew; being driven out in large quantities into the veins; which are therefore very numerous as well as very tumid. The existence also of an alternate constrictive motion, is proved by the connexion of the parts with the peritonæum by means of the common tunic and the alæ vespertilionum. The latter, or alæ, are like networks intervening between the two, and they combine the several convolutions of the bent tube, as the mesentery com-

* The alæ vespertilionum also contribute powerfully to this result, by binding together the curvatures of the tube, and attaching it to the front part of the ovarium.

bines the intestines. It is evident that the peritonæum is expanded and contracted by means of the muscles of the abdomen, and this, in the times of the respiration: consequently so also are all those appendages to which this general tunic of the abdomen is continued.

179. *Thus whilst the tube is alternately expanded and constricted, and this, by means of its tunics, namely the internal or nervous, and the external, it follows, that the cavity which it encloses is also alternately extended and contracted: and as it consists of folds, is alternately unfolded, and folded up again. This is the cause that forwards the ovum from place to place, and at length towards the orifice into the uterus. This must be done by some active force, that is to say, motion, to effect the protrusion from a larger cavity into a lesser: as the wont is everywhere in the vessels and tubes of the body; for instance, in the arteries, the trachea, the bronchia, the œsophagus, and many more.* There are foldings and rugæ, which also some call valves, by which the cavity is divided: they cannot fail to implicate the ovum, to adapt themselves to it, and in all parts to afford it a kind of new bed; and at the same time to hinder it from slipping back again to the ovarium. The little ovum is most active in itself, if it be not involved everywhere in a net of the kind, and at the same time in the peculiar humor of the tube, which comes from its glands, or rather from its cavernous substance; in itself too the ovum is of the highest levity, and in its state of activity is almost devoid of weight or downward effort or gravitation: and moreover the outermost orifice does not now look downwards. Thus of necessity it is, that by its own proper force and proper effort, and also by external force, it protrudes itself towards the narrower parts of the tube. For whilst the ovum is in that state; that is to say, whilst it is to the last degree vital, and is only the spirituous essence of the sperm now coalescing to form the carina, and most distinctly moveable in its least parts, it seems also to have an elastic power: viz., it can fit and suit itself to every dimension; just as can the softer blood, that is to say, the primitive or white globules of the blood: see Part I.

180. *Beginning from the first thread of existence, the embryo seems to have three different births or deliveries. The first*

when it is excluded from the ovary, and passes into the Fallopian tube: at which time it deposits its finest secundines, namely, the corpora lutea, which like intestines embraced its middle portion. Then secondly, when having passed through the cavity of the tube, it is expelled into the uterus. Thirdly, when at last it is excluded from the uterus, and the infant is born. The divers changes that it undergoes in these its first epochs and diminutive ages, are hidden from our ken: for of this state and time sight has no experience. But thus much we know, that three natural instincts seem to possess it: namely, 1. *of detaching and loosening itself from the ovary and its bonds; and of struggling into the tube which is ready to meet it:* 2. *of desiring to be expelled from that tube, as a vas deferens, into the uterus:* 3. *of passing from the uterus into the light of day: of which last birth, and the instincts and aiding causes that accompany it, we have some little knowledge from observations.* We shall treat in another place of the last throes of these deliveries, whereby the fœtus is expelled from the uterus. But as regards the first birth, or that from the ovarium, almost a similar one occurs in all the plants, flowers and blossoms of the vegetable kingdom. There is a kind of farina or seminal aura, which excluded in its ovaries, seems to be conveyed by the adaptation of the leaves, as of the fimbriæ of the tube, into the theca where ultimately the little seeds lie concealed. This was proved by many observations, and clearly demonstrated to me by my deceased friend, Petrus Aroselius, a man of the greatest distinction in botanical science. We may conclude then that animal nature in these respects conspires entirely with the nature of the vegetable kingdom: shewing that there is a kind of most common law of nature, welling out of the innermost sanctuaries of her science.

181. *There appear to be most pregnant causes why the fimbriæ of the tubes are not constantly applied to the ovaries, but float in the abdomen, and why the orifice itself is turned downwards, but at once, as soon as the time of the first birth is at hand, the tubes orderly erect themselves. Many causes indeed occur to us, but conjectural and hypothetical; among which we may mention the caution, lest the ovaries, by the more rigid and continued contact and rubbing, and consequently the ova themselves which are under them, should be irritated. The least*

friction insinuates a kind of vibratory motion, which would be a cause of waste to the first formation. When therefore at length they approach together, and make contact, at first softly, and afterwards a little more roughly, then by the continuance of the pulsatile motion, the ovum seems to be excited to sufficient force to burst the outermost tunic, means not being wanting to assist it. It appears necessary that the ovaries should be free from contact, and as it were left to nature, acting from the inmost grounds, to prevent any change happening to them from external forces, and perturbing the natural series. Wherefore I know not whether there be anything in the whole body that is so free by one of its extremities. This extremity appears to be an end, or a kind of ultimate, by which no communication must take place with the body any longer: for something new is to be there propagated, that will enjoy a body of its own, and not be connected with the parent body. So that when the ovum bursts out from thence with the first animal form, it bursts out into a kind of different body, not common to its parent. It is also the ultimate thing that corresponds to the organs of the man. The tube and the uterus are members proper to the woman, which do not correspond with anything in the male; and they may be proper to the animal; but they rather belong to the future offspring, being designed as its beds of formation. But these matters are only likely, or perhaps probable.

182. *But the nexus of parts enables us to deduce, that the active and reactive powers succeed and go forth in the same order as conception itself. The first active force in coition begins in the neck of the womb, where the forces at the time are at once concentrated. Afterwards it exists in the vas deferens: then in the very place and cushion of the ovum, which during the first days of conception is the beginning of the whole action that exists in the blood and fibres, or in the circumferences. Then, in every point where the little ovum passes in its downward course through the Fallopian tube. Lastly, in the uterus itself, where this ovum attaches the placenta to the uterus. Then to this spot as to a kind of active beginning, the whole determination of spirits and blood suitable for the work, comes in confluent streams. The very ovum, and the appetency and lactation, of the included embryo, is what induces these changes. The quantity and quality whether of blood*

or of spirits that is anywhere expended, determines of necessity a corresponding supply from the whole mass of the blood. Thus wherever the ovum penetrates, it forms a kind of centre and as it were living source, to which the rest stand related as streams. And this seems to be the reason why the fimbriated extremity of the Fallopian tube is not attached to the ovarium. Were it attached, this centre of activity would be unable to perform this course and circle; for in that case these activities might possibly proceed in inverse order; that is to say, the blood of the tubes might even be drawn up back towards the ovaries; and further, even from the uterus itself into the tubes; when yet the procession must take place in such an order, that the activities run from the ovaries into the tubes, and in this direction onwards into the uterus. But these statements are perhaps a little too obscure. The meaning is, that the blood and spirit of the fibres on the first days of conception, is to be derived from the whole mass to the ovum and its corpus luteum or calyx; yet in such a manner as that the Fallopian tube also shall enjoy an abundant supply of blood. This would not be the case were the tube attached by these extremities to the ovarium; for then the blood and spirit of the tube would also run into the ovarium, which would cause the tube not to be properly erected. Afterwards, however, when the ovum has got into the tubes, then the ovary no longer requires this liberal supply of the purer blood or spirits, but the tube itself even draws juice from the ovary through the alæ vespertilionum. Consequently when the passage or transmission through the tube has taken place, then the afflux of blood to the uterus and in fact to the placenta comes on next: but this order of procession would be impossible were there an immediate communication between the tube and the ovary. I am not indeed unaware that one observer of the greatest weight (Winslow) affirms, that the longer fringe of the tube is attached to the ovary; but inasmuch as on the other hand there is a general consent of the anatomists to the fact, that this fimbriated or frilled extremity of the tube floats freely in the cavity of the abdomen, I cannot help founding my reasoning on their auspices. Yet I grant that what I have said may possibly appear imaginative; a matter in which I commend myself to the judgment of others. Thus much at any rate is clear truth, that the universal mass of the blood regards the ovulum, in which the live embryo is enclosed, as the centre of activity, wheresoever it be attached; and the embryo milks out something from the blood: which milking action therefore is the cause of mutations in the body of the parent: as will be seen more plainly in the uterus, where all the arteries and veins not only swell up, but completely adapt themselves in such a manner, that they aim and strive to meet in confluent streams about this new centre of life. The natural equation also requires, that the peripheries shall supply whatever the extremes consume; and the ovum, be it tarrying where it will, is still in that very extreme.

CHAPTER V.

FURTHER REMARKS UPON THE UTERUS.

183. FOLLOWING *the successive order of operations, the uterus comes last; for the uterus takes up the ovum, attaches it to its parietes, cherishes the embryo that it encloses, and nourishes it out of its own life and blood: and accommodates itself to the several offices that are required for the formation of the new mannikin, before he begins to live with his own proper life.* This is very clear from the structure of the uterus, and also from the mode of formation.

184. *As soon as ever the ovum glides into the uterus, it attaches itself to the neighboring part, where it fixes its abode; in fact applying to the parietes of the uterus the same part that it had applied before to the tortuous corpus luteum. The ovum, relatively to the determinations of the fibres of its tunic, hence relatively to the determination of its motion, when it is naturally left to itself, must be conceived as being furnished with a sort of pole or centre of rest. This centre is represented in these very points where in the first instance it was attached by its radicle to the corpus luteum in the ovary. This its radicle, when it has passed into the uterus, cannot fail to be affixed to some particular plane of the nervous, smooth, lubricous, inmost tunic of the uterus; and to accommodate itself completely thereto: of consequence therefore neither can it fail to interpose there a kind of placenta, and by this again to draw out a great peduncle or footstalk, which is the umbilical cord.* It cannot be considered doubtful by any, that in the ovulum which is otherwise round, there is some part that formerly in the ovary was attached to the corpus luteum as the primitive uterus, through which the seed together with

the foreign spirits insinuated itself. The peduncle itself is a matter of observation. And this part must of course be the most quiet in the sphere of the ovum; according to the rules upon which we have constantly dwelt above. The direction of the fibres, and following the direction, the nisus of circumvolution, must stand related to this place as to a pole or focus. Thus when this ovum was torn away from its first womb, there must necessarily be a little scar or cicatricula left at the spot, and a thickened part, formed of the retracted extremities of the original peduncle. Thus when the ovum rolls round and round, whether in the Fallopian tube or in the uterus, it cannot turn to the obverse inmost membrane of the tube and uterus any other part but naturally this; for through this alone the natural way inwards exists or is possible; and likewise in the same part is the centre of rest, consequently the centre for the influxion of new aliment. *To this exactly corresponds the inmost tunic of the uterus, which is smooth, and consists of nervous fibrils, and applies itself homogeneously to the tunic of the ovum, which consists of similar fibrils. The ovum cannot therefore be applied to this tunic, until the former has reached such a dimension in its growth, that the instrumental causes, which are the fibres and vessels of the uterus, correspond to it as the principal cause. For this reason it is, that it must make some little delay, and a kind of journey in the Fallopian tubes; but also on its way must here and there have struck its root: and thus inaugurated by continual usage in the tubes, it maintains for itself this faculty [of taking root] entire all the way into the uterus.* From this it is easy to see and conclude, why the ovulum passes through the Fallopian tube; viz., in order that in the several points that it passes over, it may contract a slight delicate beginning of adhesion to the parietes and folds: nor is it difficult to see why the Fallopian tubes and the uterus are furnished with a subtle nervous coat.

185. *This ovum then, fecundated with the rudiments of the new embryo, establishes a kind of centre of blood-vessels and fibres in the place where it has struck its little root and fixed its abode; for whatever the uterus possesses in all its borders or peripheries, it brings together hither, and the several streams dispose themselves as circumferences relatively to their centre of activity.*

Wherefore with one common eye now look hither whatever portions there are in the dense tunic of the uterus, likewise whatever portions are seated and grouped in the substancc of the Fallopian tubes as well as of the ovary. If then all these fluids aim hither, it is necessary that as it were new ways, blind ways before, should now be opened in the direction hither; and that all the ducts, or as it were paths of determinations, should be brought together hither. Thus whatever the vascular substance contains, and whatever the common tunic continued from the peritonæum contains, and whatever the innermost or nervous tunic contains, must follow this bidding. The latter, the nervous coat, which plays the first part, and brings to the spot the purest juice, that is to say, the spirit, seems to be bestowed hither with all its fibres, and to converge into the placenta; in order that the fibres of which it is composed, may be continued quite into the new body, and following the umbilical cord, that is to say, in the body itself, may combine and actually unite with the fibres proper to the embryo; even until the proper fibre of the embryo, hatched from its veriest brains, can itself play its own part and plead its own cause, apart from any assistance from the fibre of the mother. Thus at this time the whole bodylet of the embryo may be considered as an appendage to the body of the mother, and this, until the connexion, thread or umbilical cord is broken; or until the two brains of the embryo have been carried to such an age, that it can provide itself both with fibre, and with juice and blood. The fact that the innermost membrane of the uterus is derived by contraction to this centre, is evident from its loss in pregnant women.

186. *From this concentration it follows as a matter of course, that in the spot where the placenta is fixed, a peculiar corrugation is found; and as there is no want of blood-vessels of both kinds, or of nervous fibres, therefore in the proportion of the activity, according to the common custom of animal nature, a species of muscle is there formed, whose fibres adapt themselves completely to the ratio of both the motions.* Ruysch makes mention of a kind of orbicular muscle in the part where the placenta lies: many of the keenest anatomists however have sought for this muscle in vain; instead of which they have only found fibres resembling motive fibres determined to all directions separately. But nothing is more common in the animal kingdom, than that

wherever any motion or reciprocal action exists, a muscular fibre should be formed; nor is anything more easy than the formation, since an abundance of vessels and an abundance of fibres are present; and it is of these that muscular fibre is composed. This operation comes in play in the case of various preternatural growths; how much more in this, which is to the highest degree natural. The fact is also confirmed by the motion of the uterus after delivery; a motion which is sometimes violent; and of which Ruysch makes mention; when the fibre, already ruptured, contracts in various modes, and still, as if it would continue its activity, retains for a good while the pristine motions, and endeavors to act likewise. In this case of course all the fibres and every vessel that has made its way thither, must play a kind of common part.

187. *The second office of the uterus is, to foster blandly the embryo which it encloses, and to feed it with its mother's life and blood as its months roll on. In order to its right performance of this office, not only does the uterus so dispose itself, that what it carries or gestates in the bosom of its tunics, it as it were brings together hither, but it also does the like for whatsoever the universal body of the mother possesses in the vessels and in the fibres. For the novel infant or embryo now naturally considers the whole surrounding body of the mother as a microcosm devoted to it, and all that the mother has all over, the infant regards as its own property. Indeed the mother seems at this time to live for the new infant, but not the infant for the mother.* A similar relationship is evident after birth. The maternal love is a union of such a nature, that the mother seems to live for her offspring, and not for herself. The soul of the mother appears to wish to transplant itself into the new soul of the offspring. This is the cause of the maternal storge or passion, and also of the effect when mothers encounter death for their infants. The like also obtains in brutes; the most timid animals lay aside all their fear at such times, and advance against, and fight with, their enemies; and in fact go to meet their fate in order to save their progeny. The like therefore occurs in the uterus, though it does not arrive so distinctly at our consciousness, for it lies in the occult sphere; nor is that affection excited by any external sense, either by sight or touch. In animal life however it is

requisite that the external senses also should be present, in order that the whole series may be completed. As long as any effort lies in the soul alone, it lurks under the appearance of instinct; nor is taken cognizance of by our rational mind, until similar causes in its posterior and sensible world contribute their parts.

188. *In order then that the enclosed fœtus may draw from the body of the mother as it were from the sphere of its own microcosm, all that the plan of formation requires, from the very beginning to the ultimate end, the blood-vessels, not only in the uterus, Fallopian tubes and ovaries, but also, in conjunction with these, all the blood-vessels throughout the body, must constantly intercommunicate; so that there shall be no freightage carried by any little artery in any part, that is not common to the vessels of the placenta, consequently to the body of the embryo.* This is the reason why all the vessels of the uterus so constantly communicate with each other; that is to say, both those which come from the hypochondriac vessels, and those which come from the hæmorrhoidal; and also those that arise from the spermatic vessels. Thus when any single artery is inflated, all are inflated. And this is the reason of the very frequent anastomoses and of the singular twirls which occur in the vessels; viz., because thus all the vessels are enabled to apply themselves to the centre of activity, and to be confluent in that direction. These hypochondriac or internal iliac vessels spring from the very bifurcation of the aorta; while on the other hand the spermatic vessels come off from the aorta itself close under the emulgents. These two divers origins enable us to infer, that there is not one drop of blood that passes through the heart, and is propelled into the aorta, but is common to the uterus, that is to say, to the infant which it contains.

189. *Yea, and the particular freightage of every fibre whatever that belongs to the body and springs from the cerebellum, is also common to the embryo; and on this account the embryo can extract from the very fountain of the cerebellum just as large a quantity of spirits as it requires.* This is clearly shewn in the determination of the fibres. All that mass of fibre that at length issues in the sacrum from the spinal sheath, is the residue of the cerebellum after the passage through the spinal axis. All the fibre too that constitutes the mesenteric plexuses,

is the fibre of the cerebellum; for those plexuses are made out of the intercostal nerve and the eighth pair of nerves; and these nerves are proper to the cerebellum, as we shewed in Part I. I find that this was a fact known to Hippocrates. This fibre, at this time, is in the fullest stretch of its expansion and at the top of its prodigality, for it opens itself right out to supply the formative substance: see above. But fibre does not communicate in the body with fibre in the same manner as vessel with vessel, (for the fibres run in single canal from their beginnings to their ends in the muscles,) and hence this effect is referable to the fountain-head, which is the cerebellum, where the communication does take place; and whatever fibre, at its other end, makes increased demands, supplies a larger quantity from its other end, or source. Hence upon the spring in the cerebellum depends the circumstance, that all the forces of the spirits make common effort hither, and leave the rest of the fibres no more than what is superabundant, and not requisite and necessary for the infant. These facts are the causes of a great number of variations and symptoms that take place in pregnant females; and of which we shall speak more presently.

190. *Besides the vessels and the fibres, the whole interior surface also, or inner tunic of the mother, namely, the peritonæum, and in part the pleura also, pours on all that it has in its fibre, and in its cellular tissue, agreeably to every call which is made by the progressively forming infant: and thus nothing whatever is wanting to its radical nutriment and increase, so long as anything remains in the mother's body from which it can draw forth the essences that it requires.* The peritoneal tunic with its cellular tissue is continued all the way from the bladder to the rectum, and folds over upon the uterus; nay, it does not run by a simple course from the bladder around the uterus to the rectum, but it also passes by the several ligaments of the uterus, the broad ligaments and the round,—it passes around the Fallopian tubes, around the ovaries, around the fascicles of the vessels and the nerves, that is to say, around the spermatic cord; nay further, it is conducted by the alæ vespertilionum between the flexures of the Fallopian tubes, and between the latter and the ovaries: and in short we may say that the communication, apportation and exportation are universal: for this

process of the peritonæum terminates at last over the ossa pubis; and a part of it also in the adipose tunic, where it vanishes or ceases. And if the communication with the peritonæum be total, some partial communication there also is with the pleura; for these two membranes communicate with each other through the diaphragm. The reader may see above, that this cellular tissue transmits a species of fluid; and furthermore, that it does the like in the parts where it is infarcted with fat; particularly about the bladder: see above. Thus there is no deficiency of blood for the spirits, nor yet of spirit, nor yet of aqueous and oleaginous liquid, of every degree and species. The vessels carry the blood; the fibre carries the spirit; the peritonæum, the corresponding serosity. And by all these ways there can be no defect of the supplies so long as anything remains to be supplied in the motherly microcosm.

191. *And not only the due quantity of nutritious juice, but also the due quality, is brought to the infant by the body and womb of the mother; and indeed a different quality every instant from the first moments of conception to the last, when the embryo at length emerges from its jail.* In the earliest period it requires nothing but the purest essence and the very spirit. Afterwards successively it wants the grosser essence also, and at last the pure blood and the pure lymph. The purest and the first organs demand a substance like themselves in purity of nature; and no other. And in this way the organs make a succession of demands down to the last moments. In the meantime, the embryonic cerebrum itself begins to furnish its own supplies: for the organs of the cerebrum conceive a pure essence on their own account, appropriate it, exclude it into the fibres; and likewise the infant feeds upon the liquor amnii. Thus at the end of fœtal life so great a quantity is not required from the mother's store. This now is the cause of the many changes that exist in the body of the mother; changes, that is to say, in the temper and the body, in the appetites, in hunger and thirst, and so forth, all of which are observed in gravid women. For when the purest essence is drawn off, its inner faculty is affected, and suffers from the loss of it, and a kind of lassitude is felt in the joints, attended with an intimate sense of something wrong, of which the cause is unknown. Afterwards the middle or

proximately lower essence, namely, the animal spirit, is subtracted; and when this happens partially to the viscera, nausea, vomiting, and other phenomena of the kind shew themselves. Then thirdly, when the nutritious juice and the blood are drawn upon, irrepressible and unusual appetite is the result. In the last place, the face and the features take on a deformed expression. Thus these various drainings and exsuctions cause various changes in the mother, which changes occur in successive order.

192. *At the beginning of the time, therefore, the spermatic vessels chiefly are at work, which vessels supply the purest blood, and communicate continually with the hypogastric vessels of the uterus. Next, the spermatics holding back, the hypogastric or iliac vessels come in play; and lastly, the hæmorrhoidal vessels are roused. According to the quantity and quality that is required, the vessels put themselves in a state of supply, and engender new anastomoses: nay, the pores and ends of the vessels adapt themselves, and open in correspondence with the want.* It is clear from the existence of the menses not less than from the structure of the parts, that the arterial pores and extremities stand open into the uterus: also that the ducts and continuations from the arteries which were previously in a manner closed and colorless, open now and carry red blood. This is admitted; as also that the vessels, which before were of small calibre, are very much swollen at this time.

193. *Likewise also the venous vessels, which form as it were entire sinuses, and are asylums and receivers for the blood; consequently, organs of disposition—dispositories—whereby the proper quantity and quality of blood and serum is dispensed to suit every necessity of the case. And this is the reason why the thickness of the uterine substance is permanently the same throughout the growth and increment of the embryonic body.* This is abundantly evident from the observations of the anatomists. When the arterial and venous threads, previously almost closed, are again opened, and the larger vessels are expanded, the result is a softness of the parts, and a maintenance of even thickness. The dimension of the vessels supplies the decrease. The vessels then are thrown open to suit all the proportions of the uterine expansion. Of course then the like thickness of its substance

is presented all along with its increase of size. For this reason some have supposed that the uterus is a purely vascular and pulpy structure (see Boerhaave); although indeed there is an admixture of nervous fibres, as well as of ducts from the peritonæum: not to mention also motive fibres, which are only seen however in perfection in pregnant women, and after delivery disappear.

194. *In order therefore that no alien quality, or excessive quantity, may rush in uninvited, but simply what is due and requisite may be supplied to this sacred abode of the novel embryo—in order to this effect, the arteries and veins form infinite inosculations and anastomoses with each other; and capacious receptacles and asylums, as it were sinuses, are formed. The latter also open into the cavity of the uterus, in order that all the refuse portions that come down into the uterus, may be absorbed everywhere, and that this hospice may be continually and thoroughly expurgated from all external annoyances of the kind. With this view divers means of discharge are provided; that is to say, into those great veins, the internal iliacs; also into the liver, by the internal hæmorrhoidal veins, and so by the mesenteric veins; likewise into the pelvis and the cellular tissue of the peritonæum, in this lowest region of the trunk. And by this means all the residue secreted from the arterial blood is sent out distinctly by its own appropriate passages.* Thus every case that can happen, is met: and to ensure this result, the veins communicate with each other as well as the arteries, and the number and capacity of the veins is proportionably great. Experience also shews that a large amount of serosities is derived towards, and passes into, the uterus itself, from the extremities of the arteries: and this fact is further proved by the changes of state in body and mind that occur in the pregnant mother, and by the increasing motions, and inverted positions, of the embryo. At this time there is no outlet through the cervix and internal orifice of the uterus, this being entirely closed; nor is it probable that there is under these circumstances any transpiration through the Fallopian tubes: in which case the only channel open to this refuse sweat, appears to lie in the venous sinuses or the veins themselves, which stand wide open upon the uterus, according to the experience of Heister and other anatomists. For this reason

the veins are more in number, and larger in dimension, than the arteries. The precise means however whereby the venous blood is excreted by distinct passages according to the differences of quality, is a point that has not yet been unravelled. It is plain that the spermatic chord also has similar venous ramifications, and that these export the residual portion towards the emulgent vessels, as in the male sex; likewise that the hypogastric veins lie in this situation. It is also clear that the round ligament is a kind of continuation of the spermatic vessels, and is furnished with innumerable vessels, in a coat enclosed within the peritoneal tunic; and that it passes through the rings of the abdominal muscles towards the pelvis, and affixes itself thereto, viz., to the fatty membrane under the clitoris. Nor is it less certain, that the blood of the hypogastric ramifications brings the stream back into the bifurcation of the vena cava; and likewise that branches from the hæmorrhoidal veins return it into the liver where it is again fined down. Yet, we repeat, it has not been found out how all these streams are carried back, correspondingly to the diverse qualities of the blood. It is not however to be doubted, but that the distribution is so perfect, that every branch exports blood of its own particular quality: for they all intercommunicate; nay the veins of the ovaries communicate with those of the Fallopian tubes; and the veins of both, with those of the uterus; and these vessels, the while, form marvellous turns and twirls. There is also a similar unloading and discharge of lymph from the peritonæum; as is evident from the insertion of the broad ligaments at the pelvic bones, and of the round ligaments in the adipose tissue in the same situation; also, from the fact, that the outermost coat of the uterus communicates with the coats of the before-mentioned ligaments. *A similar application to that in the uterus, may be formed in the Fallopian tubes, and even in the abdomen itself; and hence we may have tubal and abdominal fœtuses; for there is a kind of absolute communion of all parts with the uterine vessels. These applications of the fœtus however cannot be attended with the same progress and success, as those that take place in the true home or native country itself, we mean, the uterus, which is exactly fashioned for the purpose of receiving the attachments of the embryo.*

195. *The uterus accommodates itself to all the offices that are requisite for the formation of the new mannikin, before he begins to live with his own life, spirit and blood. This is evident not only from the change of state which comes over the blood-vessels of both descriptions, and from the softness, expansile faculty, and constantly similar thickness that result from this change; but the same thing is clear from the position of the uterus between the bladder and the rectum, and above the pelvis, which on this account is very ample in the female sex, as also are the ossa innominata, and the other neighboring parts of this region. By virtue of these provisions, the embryo resides with the utmost safety in this abode, and moreover has the opportunity of expanding as it requires, and of growing up until its body attains the proper dimension.* Thus all the external parts are adapted to the end by their very nature, and the uterus is laid up in the safest place, between the bladder and the rectum; which some persons have taken as a handle for their doctrine, that we are born *inter urinam et stercus.* We have already treated in this chapter of the application or adaptation of the vessels, and of the expansion and softness of the uterine substance that is the result of that application.

196. *Moreover on one side the uterus is perfectly free, and can be determined any whither in the abdomen, according as its expansion leads it. The ligaments, which are four in number, namely, the round and the broad ligaments, are no hindrances to this motion, but favor it in every way; for they are clad in a covering continued from the peritonæum; and the peritonæum itself, together with all its appendages, is capable of being stretched to a very considerable degree by means of the cellular tissue which it possesses: this is perfectly evident in fattened animals. Again, the connexion with the broad ligament is so loose, that when the uterus swells up, this ligament is rather relaxed than constricted. And the same may be said of the ligamentum teres or round ligament, which is hardly stretched to any unusual degree. The ovaries and Fallopian tubes completely yield, and come nearer to the uterus, and thus make room in the abdomen.* Ocular experience proves that the expansion is directed forwards and a little upwards, where the uterus is free. The broad ligaments and alæ vespertilionum yield, but are not constricted. The round liga-

ments are the only ones that regulate and limit the position of the uterus in the abdomen; for presently they run out through the rings of the abdominal muscles, and are attached to the adipose tunic; and hence they are placed in the corners of the fundus. Nevertheless they relax; for they are enveloped in the beginnings of the broad ligaments, as well as of the ligaments of the ovaries, and they form a duplicature in this situation, which at the time of pregnancy appears to be obliterated. Means are thus taken to prevent any of the viscera by any kind of pressure or leaning, from troubling the uterus; and the due rights of all the parts are maintained.

197. *In order to the right performances of these offices, the uterus is seen to have a threefold motion. It has two ordinary motions, namely, the pulsatile motion of the arteries, and the expansive and constrictive motion of the abdomen; and besides these, an extraordinary motion, which arises from the different changes of the embryo, by means of the muscles of the uterus and placenta.* It cannot be denied that it has a pulsatile motion, when so large an abundance of arteries and of blood meets within it. And that it has an alternating expansile motion, is plain from the continuous connexion of its common coat with the peritonæum, and with the several processes thereof; also from the insertion of its round ligaments into the rings formed by the oblique muscles of the abdomen. For during every respiratory act the muscles alternately dilate and constrict the abdomen, and this effect cannot but be communicated to the uterus. The extraordinary motions also are perceptible when the hand is laid upon the abdomen of the pregnant mother. Ruysch mentions a similar motion after birth: it is certain, from the connexion of the parts, that this motion is communicated, through the muscles, to the substance of the uterus.

198. *In order that all may happen in prosperous succession, the uterus is diligently guarded from every external injury, nay also from the inconstancies of the lower world, and particularly from the air, and its inconstant forces. Thus it is provided that no external circumstances shall burst in, to interfere with or derange this most noble work; but that all things, coming forth from the innermost sphere, shall succeed each other in the order of nature; and shall be subordinated each to each, during the*

process of their common arrangement or coördination. And for this reason the cervix uteri, and the orifice which terminates it, viz., the internal orifice, is completely closed up by the assistance of the whole fabric of the uterus, as well as by a peculiar viscid humor of its own. The structure itself affords some demonstration of this, in that during pregnancy all the vessels and all the forces are concentrated towards the uterine placenta. The fact itself however is more evident still, for when the orifice is closed the cervix uteri also is closed; the communication between the valvulæ conniventes and folds being of such a nature, that when the circular and oblique valves are constricted, the longitudinal, which serve as a kind of peduncles and trunks to the circular, are constricted likewise. Moreover the cervix uteri and internal orifice are furnished besides with a number of corpuscular bodies; and these all serve as so many little springs from which moisture oozes: they are termed, indeed, glands, vesicles, hydatids, and have been described by many anatomists. The humor which they secrete, and which is of a kind of glutinous quality, completely blocks up these cavities thus constricted of themselves; as experience proves.

199. *This same blockade of the lower part of the matrix, is a careful provision against the possible existence of any new conception, or superfætation. By means of the closing of the cervix, and at the same time the expansion of the uterus, the whole passage that leads from the cervix to the ovaries is entirely closed.* When the internal orifice of the uterus and the tube of the cervix are closed, there is no longer any passage open into the vesiculæ seminales which lie there in groups, and which open when the tube of the cervix opens; see above. The vasa deferentia or ligaments of the ovaries are also constricted by the expansion of the uterus; for the duplicature of the round ligaments is opened out, and thus the vas deferens is rendered useless; consequently the ovaries cannot be impregnated, but are retracted towards the body of the uterus. Nor can the Fallopian tubes be any longer applied to them: nay, they can undergo neither their expansion nor their inflammation (*i. e.*, erection); for all the stream of determination, both in the blood and the spirits, is now directed towards the placenta uteri. In this manner the natural order, such as it ought to be

when conception takes place, is quite changed: see above. *With regard to further considerations respecting these and other offices, we shall be enabled better to deliver them when we have treated of the progress of the embryonic formation; that is to say, in the chapters on the placenta, umbilical cord, amnios and chorion, or on the secundines, and the act of birth: the uterus being formed for the sake of all these parts.*

200. *All the differences whatever that we have mentioned above as subsisting between the female and male sex spring from a single end and principle; namely, from this, that the female shall represent the passive force, and the male the active; and that the two forces together shall constitute a single being, and by their union become the progenitors of a similar third. This is the very principle or beginning, consequently the very end, of the matter. It is from this source that all those varieties that occur in both sexes, derive their origin. As is the end, so are the means, and such are the effects, which latter completely correspond to the ends. Therefore, as is the soul, and the rational mind, and as is the animus, so also is the external form of the body. This is why we know by the face and eyes what is going on in the depths of the character and mind.* These statements however are psychological: all persons acknowledge their truth; and therefore we shall not dwell upon them. We demonstrated above that the female represents the passive force, or is in a manner, form; and that the man represents the active force, and is in a manner, essence.

201. *From this principle, and this end, it follows, that the fibres of females are less active and more passive; and particularly so, the organs which are the principles of the fibres: their fibres consequently are more tender, more soft, ready for every change of animus; more lax also, and more prone and clever in the transmission of their spirit. Indeed, the fibre is governed by its animal spirit; and as is the spirit, such is the animus; or rather, as is the animus, such is the fibre; and as is the fibre, such is the whole organism of the cerebrum and the body, which consists entirely of fibres and blood-vessels, the latter of which again are fibres in the last degree of composition. Therefore, as is the fibre, such is the whole organism of the body. As then is the organism of the body, such is the outer surface, and the counte-*

nance. Hence in that sex which is to bear the palm of tenderness, patience and beauty, the beauty itself is a speaking sign that the sex wears a passive faculty, and not an active; which latter represents a hard and resisting quality of form. All these positions flow as consequences and conclusions from this single principle. We shall have to shew in the Part on the Fibre, that in the body there is nothing that determines the real form but the fibre.

202. *Such being the condition of the woman relatively to the man, it follows that she is endowed with recipient, passive, and at the same time reactive organs. What is passive must also be reactive, otherwise there is no limit to the activity, nor can the thing exist as a substance in itself. Out of these first and middle causes proceed those innumerable details of difference but yet conformity that we find in the two sexes. Thus we observe, first, that women are more prone to be excited to changes by the external senses. That relatively to the violence of the emotions or affections of the animus, they have less judgment in proportion as they regard ends at a distance; and they regard the more present ends, to which emotion prompts them, as the most rational of all. They have a more lively imagination than men, and their rational mind, which is a higher kind of animus, plays chiefly a passive and reactive part, and descends, with much persuasion, into the imaginary phantasies of the lower mind. It is not so in the male sex, which is stronger in the rational mind, and its active force, than in the animal mind and the reactive force; and by virtue of this, not only gives the law to the affections of its own animus, as the servants, or the forces of a lower order, but also gives the word of command to the wife or consort, who does not possess the active but the passive power of will. Whence it naturally follows, that the woman is subject and obsequious to the decisions of her husband in all the matters in which rational ends are concerned. And lastly it further follows, so far as regards the body, that the woman is very prone to the motions to which her animus is excited by external objects, and rushes at once into actions, lest her desire should be thwarted by guidance. Hence it follows again that in woman all those liquids are more easily excited that serve as means for developing any of the emotions of the animus; her bile, that is to say, is easily roused to anger and irritability; the heat of her*

blood is easily excited; also the peculiar seminal humor of the female; not to mention the other fluids; and for this reason it is that woman is richer in external excitements; for the extremes or last effects correspond to the first through the means. And hence also it is, that women arrive at maturity sooner than men, namely, at fourteen or fifteen years old, while men require a much longer period for their full development. The more the innermost human faculty prevails, the later does the maturity occur, and the longer time does it consume; and on the other hand, the more the exterior powers prevail, the quicker is the advancement. In the latter case the fibre is comparatively soon ready to supply its nutritious juice; but not so where the inmost sphere is concerned, and the late supply has to come from thence. For this cause it is that brute animals, which have no rational minds, are ripe and mature in a few years after they are born. From this same principle it arises, that women, arriving at maturity more rapidly than men, also come more rapidly into such a state, that they are bound to send off a superfluous portion of spirits and blood; the superfluous part being expended naturally upon the genital organs; and as they are excited more by the animus than by the mind, it follows that their blood is more contaminate, and affords a comparatively large material for cleansing by the menses. For the more we are governed by the purer rational mind, the purer our spirit and our blood are. Every nature begins from the inmost sphere, and passes to the extremes, or from the centres to the surfaces; whence if the blood that is on the surface be rightly disposed within, it is not so easily defiled, but it dispels heterogeneous elements by its own proper and inherent force. From these remarks we see why women are passive not only in physical acts, but also in moral acts, whereas men are in their very nature active. And on this same ground also women are more beautiful and more tender than men; as it were graces in their own passive way and character; furthermore, they are more prone and determinable in all their decisions than men; and in all surface matters they appear more intelligent. The genius of the age consists in the predominance of the imaginative faculty, and in the merely passive and reactive condition of the human rational mind relatively to the*

* It is amusing to find Swedenborg casting back the imputation of imaginative-

objects that come through the external senses. The activity, however, of that rational mind, and its resistance to the emotions of the animus, that is to say, its possession of a force that can rule and govern, is reckoned at this day no proof of genius, and scarcely of judgment; and for this cause it is, that the men cannot fail to be subject to the women, for the consent of the majority and the voice of the age are in favour of this situation of affairs.

ness upon a period which has made the same charge against himself; and substantiating his accusation by an appeal to the admitted petticoat-government throughout the world.—*Tr.*

CHAPTER VI.

THE UTERINE PLACENTA.

203. Heister. "The placenta was called hepar uterinum or the uterine liver by the old writers. In women the number of placentæ corresponds to the number of fœtuses; but nevertheless in cases of twins the twin placentæ are often grown together; and yet the vessels of the two do not appear to communicate. In many animals, however, especially cows, numerous placentæ are found, sometimes as many as a hundred for a single fœtus; in which case the placentæ are termed *cotyledons*. . . . The placenta is orbicular in shape, about eight or nine inches across, and one inch thick. Its connexions are as follow. Its convex and spongy portion adheres to the uterus by the intervention of a very fine, reticular, villous membrane, continuous with the chorion. Its concave side, on the other hand, which receives the fœtus, and displays many and ample blood-vessels, and moreover is joined to the umbilical cord, is surrounded with the polished membrane of the chorion and amnion. The part of the womb whereto the placenta adheres is not determinate or constant; for in fact it is fixed sometimes to one portion of the uterine cavity, sometimes to another. The most general place of attachment however is to the fundus or posterior part; in which situation Ruysch not long ago thought that he had discovered in puerperal women a new muscle consisting of spiral fibres. According to the majority of recent writers the substance of the uterus is glandular, being made up of an immense number of small glands; but these glands, it is admitted, are invisible. Hence Ruysch has the better opinion when he states . . . its

substance to be vascular, consisting of nothing but vessels, namely, the umbilical arteries and veins, divided in the finest way, and interwoven with each other. *(Comp. Anat.*, n. 242.) The question is agitated among anatomists whether or no there is a reciprocal circulation of the blood between the mother and the fœtus: and there are many who deny that such is the fact, or at any rate doubt it. . . . Cowper found that mercury injected into the umbilical arteries of the fœtus entered the veins of the uterus of the mother: and on the other hand, Vieussens observed that mercury injected into the carotid arteries of a pregnant bitch, not only ran down into all its members and viscera, but also penetrated to the whelps enclosed in her womb." . . . Heister relates that he himself dissected the body of a female who died of hæmorrhage. This was a case of twins, one of which was born before the hæmorrhage occurred. The other fœtus was still retained in the uterus; and "he found that it had no blood, either in its heart, or large vessels. . . . The reciprocal circulation between the mother and child is also affirmed by Rauholt, Morgagni, Salzmann, and others." *(Ibid.*, not. 36.*)

204. Th. Bartholin. "The placenta is of a circular form, but the circumference uneven; and I have observed in it five prominences at regular distances from each other, with the membrane of the chorion thicker in the intermediate spaces. Where it is turned to the uterus, the placenta is gibbous, rough, and wavy, like bread which has split in the baking; and when this convex part is cut into, it presents an infinity of fibres, which may be traced until they lead to the trunks of the veins. *(Anat.*, p. 304.) The placenta is thicker in the middle of its little eminences or hillocks; but about the extreme borders it is much thinner, as also are its capillaries, which are interwoven with a variety of texture. *(Ibid.*, p. 305.) The placenta has vessels, veins and arteries, which ramify through it, and come from the umbilical vessels; all of which gradually become smaller about the margins of the placenta, and form remarkable plexuses. . . . . I have noticed of the veins of the placenta that the contained blood could easily be driven, either with the finger, or an instrument, towards the trunk or cord, but not towards the placenta. The contrary was the case with the arteries, which most readily allowed the transmission of the blood to the hepar

uteri (placenta) when they were stroked by the finger in that direction; but not so readily admitted it to the trunk. . . . The placenta by its little veins sucks the maternal blood mediately from the veins of the uterus; prepares it, and attempers it to use. . . . A part of the blood returns to the placenta through the umbilical arteries, which come off from the iliac. *(Ibid.*, p. 307, 308.)

205. Boerhaave. "The ovum now begins by its own mass to fill the whole capacity of the uterus, to apply its convex surface to the concave surface of the latter, to join with it, to unite with it in all parts by giving forth on the one hand, and by receiving on the other, vessels of the most delicate kind; chiefly however in the situation of the little placenta, where rising to form an uneven tubercle, it insinuates itself into the ample hollows of the distended uterus. During this process the uterus also is extended in all directions, and in the like proportion all its vessels increase in their several dimensions, receive more humors, and consume the material of which the menstrual plethora consists. Hence likewise in precisely the same proportion in which the channels of the ovum increase, the emissary vessels of the uterus are dilated. *(Inst. Med.*, n. 676.) The umbilical cord consists of two vessels which conduct humor from the fœtus into the placenta, and of one larger vessel also that leads from the placenta into the fœtus. *(Ibid.*, n. 677.) But when this mutual enlargement of the mouths of the vessels has increased to such an extent, that the internal emissaries of the uterus can now send forth the blood itself, and the other vessels receive it; then the abundant blood contained in the large arteries of the uterus, and driven against the uterine placenta, seems to be impelled into its dilated vessels, to undergo an alteration in its substance, and to be transmitted to the fœtus. The hæmorrhage which takes place in abortion, in delivery, in wounds of the gravid womb, or of the womb during labor, and during separation of the placenta, seems to prove as much as this: so also does the loss of blood in the fœtus when the mother only is wounded. In the dead body however the non-appearance of the vessels, and the entire membrane surrounding the surface of the placenta contiguous to the uterus, and rendering it smooth and polished, appear to hinder the

above transmission. *(Ibid.,* n. 678.) The whole body of the placenta consists principally of innumerable arteries, spread and scattered in all directions, without any apparent glands, as in the cortex cerebri; of veins intertwined in the same elaborate meshes; also of lymphatic follicles, and of lymphatic vessels. Moreover it is surrounded by a fine membrane, easy of laceration, given off by the chorion over the convexity of the placenta, where it joins the uterus. When the recent placenta is gently squeezed, the above membrane sweats out little dewy drops of blood and other humors from an infinity of oblique oscula: the oblique vessels themselves may be seen on the membrane by the microscope. The placenta receives two arteries which revert from the end of the descending aorta, and go to the umbilicus of the fœtus, and which there passing out, and being supported and defended in the semi-cartilaginous substance of the umbilical cord, . . . carry back from the fœtus the blood which is not so fit for its circulation, and thence is driven away. The origin of the veins however is so subtle, that it is not to be discerned at all; they are ultimately collected to form one trunk, which goes to join the umbilical cord." *(Ibid.,* n. 679.)

206. SCHURIG. The experience of *Rayger* and *Louis Gayant.* In a dead fœtus they blew in air through a little tube into the umbilical vein which was previously opened for the purpose; this air was given back through an opening in the umbilical arteries; having caused intumescence of the substance of the liver in the meantime. Afterwards, for the sake of greater certainty, a blue liquid was poured in through a syringe, which in a very short time first expelled a cruor from the arteries, and afterwards came out itself. So Schurig relates. *(Embryologia Historico-Medica,* Dresd. et Lips., 1732, I., 3, § 7, p. 103.) *Craanen* remarks, that thus we see that the veins in this part perform the office of arteries, and the arteries the office of veins. (p. 104.) *Fred. Hoffmann* says: it is remarkable that liquid injected by a syringe through the umbilical arteries, returns through the umbilical vein; shewing that there are no anastomoses between the vessels of the uterus and those of the secundines and the fœtus; nor does the blood of the fœtus return again to the veins of the mother. (§ 8, p. 104.)

Ruysch concludes that the uterine blood-vessels are not united immediately with the vessels composing the placenta; but the cohesion of the two takes place by means of the intervention of a kind of membranous substance. (p. 105.) *William Cowper* says that mercury thrown into the hypogastric artery of the mother, flows back through the umbilical vein of the placenta; but when injected into the artery of the placenta, it regurgitates through the hypogastric vein of the mother. (§ 10, p. 108.) *Ruysch:* The blood-vessels constituting the uterine placenta do not unite at all with the uterine blood-vessels: nor is there any gland, however small, found in the placenta. The whole placenta is made up of vessels carrying blood, whereby the chorion and amnion, and the umbilical cord, are all connected to each other. To these add the membranous muscle which has lately been discovered by me, and which is connected to the inner surface of the placenta, which is affixed to the uterus. The whole body (of the placenta) consists of nothing but vessels, the ends of which go to form little vascular brushes. (§ 15, p. 113 *s.)* The placenta is easily lacerated on that surface which is turned to the uterus; and when lacerated it seems to divide into various little globes. Ruysch concludes from his wax injections, that the uterine blood-vessels do not immediately unite with the vessels of the placenta, but that the cohesion of the two is effected by means of the before-mentioned membranous substance. (§ 16, p. 115.) The peculiar membrane that covers the placenta is called by Ruysch the juicy (succosa) membrane. So far Ruysch. A number of fibrillæ of some considerable size come off from the chorion, and are implanted like roots in the uterine tubules. (§ 17, p. 115 *s.*)

207. Ruysch. "In pregnant women the abdomen is distended to a great size, and this is attended with difficulty of breathing, swelling of the feet, anorexia of the stomach, anxiety in the præcordia, swoon, and pallor of the face. The placenta remains in the uterus not seldom for ten months or more, although the fœtus has been born. In this case placentas sometimes degenerate into hydatids, or vesicles hanging from footstalks, and filled with aqueous humor. These are not moles: although the placenta is contracted by the uterus into

different shapes, so as to resemble moles. The placenta also sometimes hardens from prolonged delay in the uterus; and Ruysch gives four figures, in one of which the placenta is all over bunches or clusters. *(Observ.* 28, p. 25 *ss.)* When the cord is inserted precisely into the centre of the placenta, for the most part it comes away with great difficulty. If however it be implanted more to either side, it easily and indeed spontaneously follows into the world after the fœtus is born." *(Obs.* 97, p. 90 *s.)*

208. In *the whole nature of the three kingdoms of the world we meet with nothing more wonderful, or more worthy of our investigation, than the formation and as it were new creation of man in the womb. It is like the existence of the microcosm in the macrocosm, or like the sending forth of a new system, as it were a new star, from its ovum and its principles.* Every individual animal kingdom may not inaptly be compared to a world. We are like the planets careering in their great vortex or universe, which latter constitutes the planetary society. The laws also of the universe are suitable to both cases. Each orb is governed by as it were its tutelar divinity, and has its own soul, to rule that particular world. But this is a comparison into the details of which we cannot fully enter.

209. *This new animated world which is coming into existence, in the course of its passage from the deeply secret places of the ovaries, through the long circuits of the Fallopian tubes, until it arrives at the fixed abode where it has decreed to pass its first life; that is to say, until it comes into the mother's womb,—attaches itself to the wall of the cavity, and exactly to that spot from which it can command a round view of the whole of the parental microcosm; and whence it can depasture and order in the freest manner all that spirit and blood that can possibly be required in the work of formation; for whatever the macrocosm has, the embryo has the power of claiming as its own.* To this point, all the blood-vessels, or arteries and veins, attended by all the blood of the heart and arteries of the mother; also all the fibre and the spirit of the whole cerebellum, come together in conflu-

ence; and from this confluence, the embryo can take what it likes, and as much as it likes: see above.

210. *In order then that it may drink in with full freedom, claim as its own, and appropriate to itself, the spirituous essences of the mother's blood, it surrounds itself from the very beginning or first thread of its existence, with a general envelope, which is not proper to itself, but of adscititious growth; we mean, the double membrane of the chorion, and the amnios likewise; and furthermore with a fluid expanse. And it applies the points of this envelope or sheath to the uterine sinus; and interposes a peculiar placenta, by means of which it extracts with wonderful power whatever it wants from its great mother, or macrocosm.* Respecting this subject we refer the reader to the sequel.

211. *From this it brings out an axis or umbilicus, which it conveys, through the most tranquil region of its sphere, all the way to the liver. There are in fact two foci, umbilici or aspides, which being brought into combination by the fluent and tortile axis, stand in mutual relationship to each other. Thence it treads the liquid path, i.e., the path of the fluids, to the first hall of the maternal world, that is to say, to the right auricle of the heart, and it goes therefrom to the innermost recess or sanctum, to the sacred and spirituous palace of the macrocosmal heaven, and to its very penetralia, that is to say, to the brains. From these again, by the path of the veins, it returns to the before-mentioned hall, or to the left beds of the heart; and then it runs, with the fibres of both heart and brain, through the common royal road of all, to the station where the way divides into two; and at this point it betakes itself back again to the axis or umbilicus; and in wreaths and windings it returns again to the placenta, and meets its beginning. And at this place a field most ample is spread without it, which determines and limits the sphere of the vortex; and this field is the membranous and fluid expanse, the first of which (i.e the membranous) consists of the chorion and amnios; and the latter is the liquor amnii. But we must now consider these positions one by one.*

212. *The uterine placenta is the means of union between the gestating mother and the enclosed fœtus. The extension of this union is by means of the tunic called the chorion, through which a wonderful circulation is carried on. The end of this circulation*

is, that the purer blood, poured through this circle and this periphery, may still have respect to the placenta as its first and last terminus, and by this means, to the innermost parts of the body of the embryo, which lies enclosed in the membranes. The outermost tunic, called the chorion, is irrigated by innumerable blood-vessels, which can come from no other source than the surface of the uterine placenta, or that tunic that surrounds the placenta, and is applied to the uterus. In fact the chorion surrounds the whole placenta, and is attached to the uterus. There is no other entrance for the blood-vessels into this tunic except by way of continuation from the same tunic that environs the placenta. The fact of the continuation is matter of experience, and all anatomists are agreed about it. The blood which irrigates the chorion can be discharged nowhere but into the vessels of the placenta; no other place for discharge is provided. A necessity springing out of the continuity and connexion of the parts, proves this sufficiently.

213. *The placenta lies in a duplicature of the chorion. Thus the blood-vessels are congregated within the placenta, and the consequence is that the vessels of the chorion pass for the most part within the same duplicature. It would seem that something similar to this, some analogue of it at any rate, existed in the veriest ovum before it was detached from the ovary, but from which it was afterwards torn away; though at this early time, instead of blood the purer, that is to say, spirituous essences, ran through it, and ran in a still more perfect manner, accomplishing almost the same sort of circle as in the present case.* The tunic of the chorion, inasmuch as it surrounds not only the anterior but also the posterior surface of the placenta, is a proof of the fact that the placenta lies in a duplicature of the chorion. And moreover it is certain, that the chorion is a double tunic, in this respect not unlike the dura mater, which has the great venous sinuses between its layers, and in the same manner receives the peculiar arteries as well as veins of the brain. Similar duplicatures of membranes are common also in other parts of the body. So long as the ovum resides in its ovary, it is beset by a kind of finely intestinal corpus luteum, as it were a most delicate womb; though whether there be any trace of a placenta is not determined: it follows however from the connexion of things, that there is some

analogue of placenta, with a petiole, in the place of the umbilical cord; also that in the earliest stage of conception there was something visible between the tunics like the first thread or initiament of the fœtus. That this very rudiment however was continued inwards, is a point that will be seen as we proceed. The tunics themselves remain; for in process of time they grow large and thick; whence it seems scarcely possible to doubt, but that something similar, though more perfect, occurs previously; and that at the time when it is not pervaded by the red blood, it is permeated notwithstanding by its prior essence or spirit, which accomplishes its circles more perfectly; as will be shewn in our Doctrine of Forms.

214. *There exists, by means of the placenta and the chorion, a communion between the mother and the embryo; but this communion is not of the same character as that which subsists between the appendages, members, or viscera, of one and the same body. The mother and fœtus are at once conjoined and separate; the mother being a substance by herself, and the embryo by itself. And whatever of blood and spirit the mother possesses, is not her own peculium or property, relatively to the embryo; which draws upon it to the precise quantity and quality that it requires. On the other hand, whatever the embryo draws from this great mother, it claims as its own peculiar property, and leaves it at the mother's disposal no longer.* The communion between the several viscera of one and the same body is of such a character, that whatever one member possesses, is common to the rest when they want it; for all the blood, wherever it goes, passes at length into the grand vena cava, and returns towards the heart. Not so however in the embryo, in which no refluxion of this kind is perceived,—none that is established upon any sure basis of observations. There are indeed some phenomena which might seem to favor the affirmative; such, for instance, as the evacution of blood from the embryo when the mother has been the subject of violent hæmorrhage. However, this is not a consequence of that fact, as I shall shew presently on many grounds.

215. *Such therefore is the union, that the embryo claims for its own whatever the mother possesses; and when it has drawn it in and drank it, claims it as utterly its own property and possession. The mother however claims nothing as her's that has*

become the embryo's, nor asserts any further right of property in it; but it passes from the power of the one into that of the other. Thus from the very first the two enjoy separate rights. This is sufficiently apparent from the fact of the insertion of the fibres, vessels and ducts of the maternal uterus into the tunic of the placenta, which freely drinks the sanguineous fluids, but has its own ways and pores so inclined and so obstructed, that no return is any longer allowed through the same ducts. This has been pretty clearly shewn by the extraordinarily fine injections of Ruysch and other anatomists; who have found that after injecting the arteries and veins of the placenta, the surrounding membrane contiguous to the uterus was nowise tinged by the color of the injection, but became more pallid and smooth than before: so that up to the present time it has not been found that a drop had escaped through the pores of that tunic. This is a very evident proof, that those blood-vessels, or those vessels purer than blood-vessels, pass through the tunic very obliquely, as in the intestines and other parts: thereby permitting entrance, but no return: as Ruysch plainly confirms where he says, that the vessels of the placenta communicate in no way with the vessels of the uterus. Thus we may conclude, that whatever of blood the embryo has drawn, it claims as its property, and does not restore to the mother. The same is proved by the changes that the mother undergoes during the time of gestation, such, for instance, as anxiety about the præcordia, anorexia, swoon, pallor of the countenance, and the like effects. Whatever then has once been let into the placenta, is given up thenceforth to the growth or vegetation of the embryo. *It is however provided that the embryo shall draw from the mother no more than she is able to replace.* Were not this the case, the one body, in its rise and birth, would prove the wasting and death of the other, and like certain animalcules, the mother would spend her own life upon her offspring.

216. *In the first days of conception, the embryo demands pure spirit from the blood and the fibres; afterwards, however, spirit mingled with grosser essences,—what we shall henceforth term the white and purer blood; lastly, blood, and a certain corresponding serosity: all according to degrees, as the body increases, and as it wants the grosser for use in combination with the purest*

juices. This is plain not merely from anatomy; namely, from the state of the ovum in its ovary; but also from the uterus, to which the placenta is attached; for the vessels successively enlarge, shewing that not only a larger quantity, but also a better quality is demanded: the purer the vessels are, the more refined is the blood that they carry; the clots and viler serosities being dissipated on the way to the capillaries. Reason is in favor of the same position; for the thicker the web becomes, the grosser are the threads, and also the grosser the fluids that permeate the threads. To the same purport also speak the phenomena in pregnant women; their relaxation, anxiety, changes of emotion, and their slight bodily ailments; all of which keep the same successive order, which run gradually from the inmost and purest sphere down to the grosser. There are organs to which the ovum is attached, of the same structure as the corpus luteum, which is the first womb of the ovum, and can only be permeated by the purest blood and its spirit; and this, both in the Fallopian tube, and lastly in the uterus.

217. *Nature everywhere in her kingdoms passes in successive order from the innermost to the outermost, or from the purest to the grosser sphere; and this is the case not only in the formation of her parts, but also in the parts that are already formed.* This is the source of the perfectness of nature's works and actions. Nature proceeds from the purest parts, which are also the most perfect, to those which are comparatively gross and imperfect; and these latter she thenceforth governs from the intimate and most perfect. The reverse is the case with human art, which goes from the outside to the inside, and stops at the threshold of its journey, in the external sphere. It is also to be observed, that when nature emulges and draws the grosser, she at the same time draws those elements that were purer once, or all that at the same time *a priori* was once included down to and in those ultimates; for the interior is contained in the next grosser matter, and the blood, which is the last of the animal humors, has in it whatever had preceded. Such is the order of the fluids: first there is the purest essence; next to this comes what is called the animal spirit, and the white blood: lastly, the red blood. When the embryo drinks the spirit, it also drinks that which is prior and purer, or the very vital essence

of all. When it drinks the blood, it at the same time drinks the spirit, and in the spirit again, the essence just mentioned; for these fluids lie one within the other, and are inseparable. They are distinct, because the prior essence can exist separate from the posterior, though the posterior cannot exist apart from the prior, inasmuch as it exists from the prior; and not *vice versâ*. Thus the first rudiments of the embryo cannot be formed of mere blood; but the formation takes place successively before it exists simultaneously; for every thing co-exists as a thing by virtue of successive constituents.

218. *In the succeeding period, however, the embryo itself conceives and excludes, according to their own degrees, the proper spirits, and at length the proper blood, from its own principles and first springs of life, in the two brains and in the two marrows. Also it nourishes itself from the store that it has in the meantime been laying up. Hence gradually the want and need of food and nutrition from the mother's blood is lessened. And in the degree in which this takes place, the communion begins to perish, and the embryo commences to live with its own life, without making use of auxiliary riches from the life of the mother that carries it. By virtue of this new relation there now advances, out of the inmost sphere in the first instance, an effort and a mystic wish, to break connexion with the mother, and to claim the right of its own disposal; and with this tendency the outermost spheres conspire; and thus the embryo now attempts to open the doors, and to desert the tunics of the chorion and amnios, as well as the placenta, which thenceforth begin to be of no farther use.* This will be demonstrated more distinctly in the sequel. The cortical glands of the brain are the purest of factories, and compose and exclude that purest of essences, the animal spirit; which they afterwards transmit through the fibres into the vessels; where by a sort of copulation with the sulphureous and saline elements of the same degree, the blood arises into being. It derives these elements partly from the mother; partly from the liquor amnii; so that it no longer stands in need of any foreign resources.

219. *It appears from these various considerations, that the first essence, which regarded as a substance is the soul, is derived from the father: and that it conceives and excludes itself accord-*

ing to the increase that takes place in the purest organs of the brain; and that all the rest is from the mother. Also that the ovum cannot in any wise be impregnated by the soul and spirit of the mother; for the fact of conception demands that that first essence shall be pure, and packed as it were in globular forms, of which it occupies the inside, whilst the animal spirit makes up the exteriors or surfaces. This is an effect which can have place in the male only, which possesses similar organs, that is to say, testicles, which extract that first essence; and epididymides, which give the essence shape and form.

220. *The uterine arteries of the mother appear to extend their pulsation or sphere of activity, no further than to the first point of contact with the placenta and its outermost membrane or chorion. Thus it is that the mother cannot be said to act upon the fœtus that she encloses; but all that she does, is, to proffer it spirits, blood, and humors, and thus to give it the opportunity of acting for itself.* This is plain from the connexion of the vessels and ducts of the uterus with the placenta; which connexion is not immediate with the vessels of the latter, but mediate through the common tunic or chorion, which in this situation is pulpy, soft and bibulous. The mother is the macrocosm, or the universal subject from which the fœtus draws all that it wants for its life. Now with regard to the influxion of universals into those things that are at once distinct and separate, it is always of such a nature only, that it simply gives power or potency; whereas other matters depend upon the recipient subject itself. On these subjects see our doctrine of order.

221. *Thus the placenta with its common membrane the chorion, applied, as it is, to the uterus, is of an epispastic quality; that is to say, exercises a natural attraction, like blister-drawing and detersive substances in general: much in the same way as the mammæ and nutrient papillæ after birth, which do not pour in the milky fluid, but proffer it to the infant, to be sucked out.* This is proved not only by the fact, that the vessels of the placenta do not communicate with the uterine vessels (according to the observations of Ruysch and other anatomists), but also by other phenomena. Were the blood poured in, then would the embryo receive whatever the mother brought to it, whether clean, or unclean; and in this case it would have no action of

its own, nor would merely emulge what would conduce to the formation of its delicate frame. And moreover, under these circumstances, its growth would not be limited to the stated period of nine months, but would be according to the mother's lavish supplies; and thus inconstancy would preside over the several points in the career of formation. It is however the embryo itself that thirsts and hungers for this alimony or food; that is to say, for a precise quantity or quality of the maternal nutriment; and all proceeds according to this natural hunger and thirst: just indeed, as after the delivery, at the teats of the nursing mother, when the same instinct is active, constant and repeated.

222. *In the meantime the mother has it in her disposition, that all shall be done in order, and this she manages by the interposition of fibres, and among the rest motive fibres, between the parietes of the uterus and the tunic of the placenta; and also by the application of the ducts, according to the force of attraction exerted by the placenta.* This appears to be the function of the fibres of the orbicularis muscle, mentioned by Ruysch. These fibres flow in any direction you please, because they flow in all; and according to Heister they admit of being interposed and otherwise disposed. By these fibres of direction then all is managed according to the wants of the infant. This is a ready and a common matter throughout nature, and especially in the nature of the animal kingdom, which constantly interposes motive fibres whenever any direction of forces is demanded.

223. *Yet is there no simply attractive force, but the force depends for its cause upon the disposition of the fibres and blood-vessels in the placenta; and the disposition of the fibres depends for its cause upon the whole infant, from the very principles of life and motion in its brains. Thus is there a continuous connexion of causes, from the first and inmost of the fœtus, to the outermost in the placenta, where the latter is applied to the uterus. This connexion may even be rendered intelligible by an anatomical exposition of the embryo's diminutive frame.* The primum mobile of every living body must as a matter of course be derived from the principles of the life of that body; consequently, from its cerebrum and cortical glands, which are the extreme ends of the blood-vessels, and the beginnings of the fibres. All things

whatever that exist organically in the body, derive their existence from the fibres. This is the case even with the blood-vessels; consequently with the motive fibres, which are made up of nervous fibres and blood-vessels. As then the beginnings of the fibres have their own life or vital motion, and feel exquisitely whatever the universal body requires; and also perpetually set up the motion of the parts of their body by alternate animations; it follows, that all that is required by this principle, runs in abundance to the call, by a kind of attractive or syringic force. See Part I. This sphere reaches throughout the body from first to last; and pervades therefore all the links of the chain, which is continued in fact to the placenta. In the placenta, the vessels are so managed, that the blood abounds to them in exact proportion to their wants and requisitions. The order prevailing in them is, that they taper off into the minutest threads, by nature's usual degrees; and in a manner correspond to the principles in the cerebrum.

224. *The sphere of the activity of the embryonic body reaches as far as the placenta; and the brain of the embryo, wherein the principles of its vital motions reside, by means of all the intermediates, and lastly, of the umbilical cord itself, corresponds in such a manner to the body of the placenta, that the latter gives whatever the embryonic frame requires. This depends entirely on the disposition of the vessels and fibres, and on the continuation of the same.* We shall presently have to treat in greater detail, of the disposition of the fibres and vessels; and we shall have to shew that the vessels require to be lessened down, and drawn out, in just series, to their least pencils and brushes, or capillary threads.

225. *But all this cannot occur without a peculiar motion of the parts in the placenta; that is to say, without the pulsation of the arteries both of the placenta and the embryo. This pulsatile motion is not the same as the mother's pulse, but it is proper to the embryo itself, arising from its principles of motion, that is to say, from the animation of its brains and the systole and diastole of its heart, which at this period are synchronous actions. For this reason it is that two arteries from the internal iliacs of the embryo, are transmitted through the umbilical cord to the placenta, and ramifying through the placenta, split into capillary*

extremities, irrigating and modifying the whole of the placental substance. We shewed in Part I., that in the fœtus the motions of the brain and heart are coincident or synchronous. But for the purpose we are now considering, the arteries, viz., the two iliacs are made use of. Without these the placenta would have no pulsatile motion; consequently, no acting force serving to advance the blood on its way, or corresponding to the motions of the brain and heart.

226. *It also seems likely, that the pulses of the arteries of the fœtus, and consequently, of the placenta, do not coincide in moments, or are not synchronous, with the pulse of the arteries of the mother or the womb. For the less any body is, the more rapid the circuits of its fluids; a fact which is proved by all the subjects of the animal kingdom.* This also brings it to pass, that the placenta is so joined to the uterine vessels, that the uterine blood does not rush into the placenta at any other moments but those in which the embryo itself demands it. Otherwise the embryo would draw no blood from the mother's supply, except when she gave it forth. *This is plainly confirmed by experience; for the infant sometimes survives several hours, nay as much as two, or even three days in the womb after the death of the mother: and meanwhile, it lives on the blood of its own placenta and liquor amnii.*

227. *In the meantime a certain communion, or a kind of common connexion between the mother and the embryo, appears to be brought about by the fibres of the two, in the placenta, and at the same time in the umbilical cord. Indeed there is a peculiar nervous membrane that invests the uterus at other times than those of pregnancy. A similar membrane is observed to be interposed between the placenta or its most external coat, and the uterus. This membrane reaches fascicle-wise inwards, all the way to the considerable roots of the umbilical vein; and consequently it extends around that vein still further, and perhaps towards the liver of the embryo. The proper fibre of the embryo comes to meet it; that namely which is continued around its two arteries from the iliacs; also along the umbilical cord; and likewise the fibre that is continued from the whole of the arterio-vascular cuticular substance of the embryonic body, around the same umbilical cord all the way to the placenta. Thus the fibres of the mother*

and embryo run towards each other and meet, from the very umbilicus, and at the several parts of the placenta; though the manner in which the union takes place, has not yet been laid bare. With respect to the fact, that a nervous and porous membrane surrounds the cavity of the uterus, we refer our readers to the anatomists, and to our Chapter on the Uterus. See also the citations prefixed to the present Chapter, where it is shewn that a similar membrane is interposed between the placental chorion and the uterus. And again we refer to our authors, for the fact that this membrane winds its way in fascicle-wise all the way to the venous vessels. See again the same citations for the further fact, that on the part of the fœtus, a peculiar nervous membrane invests the several arteries, and particularly the aorta, and its bifurcations, or the iliac vessels: and for this also, that this membrane is continued by the two arteries along the umbilicus, and ramifies round about all over in the placenta. Now hence it follows, that the fibre of the mother passes into the body of the embryo, and that the proper fibre of the embryo is continuous with the same from the umbilicus all the way into the body of the placenta; and they unite with each other in some manner on the way, but especially in the placenta.

228. *We may hence infer how the affections or remarkable emotions of the animus in the pregnant mother, can affect the infant with its little frame in so powerful a manner, that actual traces of the emotions are imprinted upon it, and thenceforth remain. For every affection of the mother's animus, induces a general change of state and a particular modification on the animal spirit, and consequently on the fibres of the whole body; hence also upon the uterine fibres, which penetrate by this way into the placenta, and into the umbilical cord, and throughout their course communicate with the proper fibres of the embryo. This fibre is distributed not only to the internal iliac vessels, and to the aorta, by means of the innermost coat of those arteries, but also by the several arteries that so abundantly irrigate the delicate cutis of the embryo, and all of which in like manner meet in the aorta; the umbilicus being continuous with the cutis, and consequently with the cuticular vessels of the infant. Thus the whole of this change of state in the spirits and fibres, coming from the entire surface of the embryo, meets on the internal tunic*

of the aorta. This inmost tunic of the aorta at length carries it through the carotids into the whole of the cortical substance of the cerebrum, cerebellum, and medulla oblongata and spinalis; in which it terminates: and as these substances are the principles of life and motion, and also the principles of all the fibres of the entire embryonic system, it follows, that there is a communication of the affections of the mother's animus, or of the state of her brain, with the brain of the embryo: and the embryo forms the members of its diminutive frame, in agreement with all the representations of its brain, i. e., with all the ideas that are therein implanted; and in this case also according to the representations and ideas of the mother, which come upon it, as though the mother poured them thither. Yet the connexion between the fibres of the mother and embryo is of such a nature, that it carefully provides against the embryo being affected by the ideas of the mother, unless those ideas be excited to the highest possible degree. Respecting the marks which the ideas of the mother imprint upon the embryo, we refer the reader to our authors. In this place I have only been anxious to indicate the mode of the communication; for certain it is that the fibre of the mother has fellowship with the fibre of the embryo: see above. The fibre of the embryo is continued from the brain of the child through all its organic parts, for all that there is of organic in the body is the domain of the fibre. The cord is continued from the skin or cutis; and also from the iliac vessels of the embryo through the arteries: this is clear. The innermost membrane of the arteries is their first and veriest nervous membrane; see Part I. This membrane is continued to the whole cortical substance of the cerebrum, cerebellum, and medulla oblongata and spinalis: see Parts I. and II. This very cortical substance is the ultimate organization continuous with these, and passes from them into the fibrillæ: see again Parts I. and II. The like state that comes upon these substances, is diffused throughout the whole system: see Part II. Now it follows as a matter of course from these positions, that the effect of the cause is presented in the exquisitely delicate framework of the embryo, which springs forth entirely from those its principles of life and motion, and from the representation thereof which lies in regarding that which is to be formed as if it were already formed: see Part I.

229. *Thus the cerebrum of the mother and the cerebrum of the embryo communicate by means of intervening fibres, and mutually correspond to each other. The influx however takes place according to the connexion that there is between the fibres of the two in the placenta and the umbilicus; which connexion seems to be too loose, to allow any but the very strongest of the mother's ideas, which are sufficient to stir the blood itself, from being represented in the brain of the child.* That the connexion is thus loose, is plain not only from the facts of the case, but also from the first connexion which there is in the womb. For the middle or interposed nervous tunic adheres only very loosely; and thus only loosely again in the inner parts or the placenta and the cord. Thus it is not very active, but rather passive and reactive; especially as it only goes to the veins. So that unless it be excited, stretched, and as it were hardened and inflamed by means of the blood, it would seem that it could not take any active part: as happens when it is thrown into a state of tension by inflammation; for every sense is put upon the stretch by the mere inflammation of its organ: this is the case with sight, hearing, touch, &c.

230. *This faculty of communication increases by degrees from the day of conception to the middle of the period of fœtal life; but after this it again decreases: and at length it gradually so diminishes, that almost no communication exists by the way of the fibres.* When the placenta augments, and the uterine vessels swell up, and the uterus undergoes a kind of inflammation, this communication also has its increments; for at this period this nervous membrane is distended and put on the stretch on both sides; and this is the reason why there is especial danger to pregnant women during the middle part of their time. But when the placenta has done increasing, and the cord is relaxed, the membrane decreases. For the connexion grows looser.

231. *By this means, in process of time, the embryo is left more and more to its own disposal, and from the inmost first, its connexion with the mother is detached: and as the innermost connexions, so also the outermost, are detached by corresponding*

degrees; until at length it has no longer any need of the sanguineous and spirituous nutrition of the mother. And then naturally it prepares and girds itself up for separation and coming forth. See above. All source and origin is from the inmost, and goes to the outmost sphere: and the outmost stands disposed to the inmost exactly as the effect to the cause and the principles of the cause.

232. *Inasmuch then as the placenta invites to itself the maternal blood and spirit by its own proper force, and at once appropriates it to itself as its own, nor lets anything pass back into the vessels of the uterus, it follows that the placenta is also a purificatory and secretory organ relatively to the blood: that is to say, that when it has drawn and emulged the blood, it purifies it, nor takes to itself any but the due and suitable portion, while it rejects the recrementitious part.* This follows as a necessity from the points already demonstrated. And that the placenta is a purificatory organ is plain from the peculiar fluxion of its vessels, that is to say, from the make of the organ. Indeed, the several vessels, both venous and arterial, split into the extremest subtleties of imperceptible threads.* And whenever nature proceeds in this order, whatever is gross and heterogeneous is rejected there: there are also seen places of discharge, and lymphatic ducts; respecting which, consult our authors.

233. *The vessels of the placenta, both arterial and venous, are split into their last subtleties; and the order of the splitting is such, that they are subdivided into exact degrees; so much so, that there is nothing that is gross and heterogeneous, but has its own peculiar part of the way where it is thrown aside, on its path to those ultimate threads. And finally, the vessels pour nothing into their corresponding veins, but what is purified of all uncleanness.* The course and progression of the arteries and veins of the placenta, have been shewn by Ruysch and others, to whose works and fine injections we refer the reader. This progression takes place by a subdivision according to degrees;

* Does it not also appear in the intellectual and moral worlds, that during periods of universal discussion, all the great subjects and problems of existence submit to this fine capillary condition, and are split into thousands of critical minutiæ, in order to be eliminated and purified, as well as in order that the old total may cease, and a new one arise?—*Tr.*

that is to say, it is a subdivision first into considerable branches; then into lesser offsets; lastly, into fine threads; and it is carried on through innumerable twirls and twists that represent vesicles and as it were glands. This is confirmed by the same injections to which we have just referred. The placenta is subdivided into lobes or prominences, and when it is detached from its connexion with the uterus, it falls into a kind of globular masses; these again into lesser globes, and these into least; almost as in the liver. This is evident from the remarkable division that it undergoes, when it is no longer a sanguineous structure, but is charged with hydatids; as for instance when it has remained in the womb several months after delivery: in which cases it has the appearance of bunches and clusters of grapes. This the reader may see in Ruysch's three figures. The distinction that we are mentioning is far more conspicuous at this time; for the organ breaks up into no foreign form, but into its own natural form, according to the divisions that existed in it before; for all that has occurred is, that the membranous ligaments, and the lesser bands, that had obliterated its divisions, or rendered them as it were continuous, have disappeared. By the progression of the arteries to their leasts, the blood is purified on the way through a course of similar degrees: see above, and Part I.

234. *And to prevent any uncleanness from creeping in by chance, and infesting the delicate organs of this little body, and in order that all that enters may be duly purified, the venous blood is delivered by the placenta through the umbilical cord to the liver, which viscus is another purificatory of the blood.* See the Chapter on the Liver, and the mode in which it purifies the blood. This is the reason why in the embryo the liver is almost the first member of the body, and by far the largest in proportion of all the viscera. Thus the placenta relatively to the veins corresponds to the liver, and relatively to the arteries to the brain.

235. *In order rightly to purify its blood, it sends back its grosser portion from the iliacs through the two umbilical arteries; and this, for the sake of equation both in the matters of quantity and quality. And for this reason the artery seems to communicate with the vein in the course of the passage through the um-*

bilical cord, and finally in the placenta itself. This is according to the observation of Cowper and several others, respecting whom we refer the reader to our authors. The quantity needed is in this way restored exactly according to requirement; the quality likewise in the placenta itself: for the fact of this same arterial blood returning by the veins, is a consequence of the circumstance, that no communication exists with the uterine vessels.

236. *Especially is the blood purified by its extensive circulation through the tunic of the chorion, which is extraordinarily rich in vessels; also in the placenta, where there are many receptacles and lymphatic ducts. Moreover, to prevent anything pure and spirituous from perishing, lymphatic vessels are added to the rest, which absorb the purer essences from the humor secreted, and bring them back again into the venous blood.* The reader will see further on, that the membrane of the chorion, unlike that of the amnios, is richly provided with vessels. This expanse is a kind of field through which the vessels can freely expatiate, and constantly bring to bear a sufficient quantity in every case and state of the mother. Respecting the ducts and lymphatic vessels that we have mentioned, we refer the reader to our authors. The lymphatic vessels everywhere absorb that pure essence that is mingled thoroughly with spirit. See our Chapter on the Thoracic Duct.

237. *But where there are secretions, there must also be excretions; for were there no excretory passages, the pituitary and excrementitious humor would stagnate there, and would besmear all over the channels of the purer blood, and entirely put a stop to the work of formation. The channels of this excretion run through the chorion into the cavity of the uterus; or into the parietes, where the largest venous mouths open, and absorb this unclean portion, and bring back the serum in its place in the liver.* Many circumstances make it plain that excretion takes place through the chorion into the cavity and veins of the uterus. This is proved by the tunic of the chorion itself, which when it is compressed between the fingers, gives out a liquid ooze or exudation: by the abundance of the arteries in the same place; by the amnios, which strictly prohibits excretion from taking place in the other direction, or inwards. Also, by the veins of

the uterus, which open into the cavity by an abundance of mouths. Likewise by the application of the chorion to the concave parietes of the uterus as pregnancy advances; which shews that at this time these parietes seem immediately to take up the fluid. Furthermore by the veins of the uterus being without valves, and communicating with all the veins, even with the internal hæmorrhoidal, and by means of these with the mesenteric veins, which go to the liver. Thus much indeed cannot be considered doubtful, that the veins of the uterus do absorb portions of the kind, and purify the uterine cavity. There is a further question, viz., whether the Fallopian tubes also imbibe any portion, and send it away into the cavity: some affirm that this is the case, and deduce it from the recrementitious matter found in the abdomen.

CHAPTER VII.

THE CHORION, AMNIOS, AND LIQUOR AMNII.

238. Heister. "There are membranes enclosing the fœtus in the womb, as it were in an egg. The outer membrane is termed the chorion, and is thick, spongy, villous, and replete with blood-vessels. It is contiguous to the uterus, to which also it is connected by filaments; and it may be divided into two lamellæ, the inner of which is regarded by some in the light of an allantois. The inner or second membrane is the amnios; it is a thin transparent membrane, contiguous to the chorion; and unlike the latter we see in it no vessels, or almost none. It holds that glutinous and pellucid liquor in which the fœtus is contained. The membranes now described are ruptured in childbirth, and let out their liquor. The allantois, or membrana farciminalis, is finely seen in some animals, and especially in cows, and is continuous with the pervious urachus. It does not exist however in man. In the aforesaid animals it serves as a receptacle for the urine. In the cow it is about twelve feet long, and when inflated with air, it is more than a foot in diameter. It is then a big and we may say stupendous part. . . . In the human fœtus the urachus is generally found closed, besides which, grave reasons may be given why God has chosen to exempt women from carrying about a load of urine in the allantois. . . . . Some authors pretend that the allantois is destitute of blood-vessels, but I have both found, and figured, an abundance of them. *(Comp. Anat.*, n. 241.) In the calf, the sheep, &c., the urachus is a pyramidal canal, reaches in the first instance from the fundus of the bladder to the umbilicus, through which it passes; and afterwards it gradually enlarges,

and terminates in the allantois, (almost at right angles on both sides, or transversely to one side,) to which it conveys the urine from the bladder. In the human fœtus it is not pervious throughout; or at least very seldom so; but is for the most part found solid like a ligament. Hence it would appear that the canal in this case can hardly perform the above function; particularly as no allantois, or other part of the same kind, in which to discharge itself, can be clearly pointed out, or at any rate has yet been properly demonstrated." *(Ibid.,* n. 243.)

239. BARTHOLIN. "The membranes about the human fœtus are two in number; about the fœtuses of brutes there are three. The above two, each cohering and growing to the other, so as to present the appearance of but one membrane, form what is called the secundine. *(Anat.,* p. 301.) The first membrane is the amnios, which on account of its softness and delicacy is by some called the lamb-like membrane, by others the charta virginea (virgin's paper), the indusium or clothing, &c. It is the thinnest of the two membranes, white, soft, transparent, and furnished with rare and minute veins and arteries. . . . It is coherent with the chorion almost all over, particularly at its borders; and united to it in the neighborhood of the placenta, and in the middle of the same, where the umbilical vessels come forth from it. We have however easily separated it from the chorion. A large quantity of humor is found in it. . . . In the human fœtus the bladder is for the most part distended and tumid; in brute fœtuses, empty. *(Ibid.,* p. 301, 302.) The second membrane is called chorion, because like a circle it surrounds the fœtus. . . . It is separated from the placenta with difficulty. . . . Towards the fœtus it is smooth and polished; but it is rougher on the side where it is laid over and agglutinated to the placenta. In the human subject this membrane adheres to the uterus mediately, by means of a round and reddish fleshy mass, connected only to the other, and this, almost always the upper and anterior part of the uterus. *(Ibid.,* p. 303.) A part of the amnios occasionally adheres to the head of the child during birth, and such an infant is then termed galeated—crested or helmetted. *(Ibid.)* The allantois does not entirely surround the fœtus, but is like a belt or sausage (farcimen)." *(Ibid.,* p. 308.)

240. SCHURIG. "*Ruysch* presents a portion of the third or

innermost tunic of the impregnated womb of a sheep, through which myriads of unknown vessels are disseminated, in which vessels the nutriment intended for the fœtus is prepared. These vessels are called vermicular from the circumstance, that they creep with worm-like writhes through the whole of the inmost tunic of the uterus. Lister however is of a contrary opinion to Ruysch. (*Embryologia Historico-Medica*, 1732, 1, § 10, p. 13.) *De Graaf:* We have found a liquor alike in color and taste, in the stomach of the embryo, and in the amnios: he argues that it has passed into the stomach through the mouth. (§ 12, p. 16.) *Heister* supposes that the fœtus is nourished by the mouth from the liquor amnii. He has proved it by the experiment of a fœtus of a calf within frozen liquor amnii; in which a piece of ice as thick as a finger was found in the mouth, and extending down into the œsophagus and stomach. (§ 16, p. 26 *s.*) *De Graaf:* That the fœtus is nourished by the mouth, is readily proved by the contents of the fœtal stomach and intestines. We have always observed in the œsophagus and stomach of the embryo a liquor similar to that in which the embryo swam: in the duodenum also; excepting that on account of the admixture of bile the liquor had a slight greenish tinge; and the further it went down the tube, the more inspissated or thickened it became. And that nutrition is effected by the umbilical cord, is shewn by the case of a whelp which was born without any head or mouth, and perished as soon as the umbilical band was broken. (p. 29 *s.*) *Stockhammer:* Six or seven weeks after conception there appears the delineation of the head, very large in size relatively to the rest of the little body; also the spine curved like a keel, and the neck no thicker than a thread; the whole of the little machine being scarcely bigger than an emmet. In the head or face there appear instead of the eyes two black points, and instead of the nose a white line. The cerebrum is like a kind of mucus, and the little skull like a membrane of leather. In the chest and abdomen the viscera are scarcely discernible. The umbilical vessels also consist of short and indistinct fibrillæ; and from the chorion or outer membrane of the ovum there rises a sort of fleshy protuberance, adherent to one or another part of the uterus. (I., 2, § 1, p. 30 *s.*) *Bohn* says that in the first months of gestation the chorion and amnios are at some distance from each other, and are kept apart by a humor which circulates between them; as may be seen in cases of abortion: afterwards they are more closely united, and look like one single tunic. *Drelincourt* and *Blasius*, in reckoning fictions upon this subject, number the space between the amnios and chorion among them. (I., 2, § 6, p. 34.) *Verheyen* states the contrary; for he will have it that the humor sweats through the pores of the chorion into the cavity of the amnios. (I., 2, § 7, p. 36.) The galea, crest or helmet, which there is in male infants, and the fillet which there is in female, upon the head, is thought to bring fortune; is also used for witchcraft: and the midwives are wont to prophesy from it. (§ 9 *s*, p. 38 *ss*.) *The liquor amnii:* This they derive either from vapors transuding from the body of the fœtus, or from sweat, or from urine. (§ 12, p. 46.) Others derive it from the brain, the nerves, the glands of the eyes, ears, jaws, mouth; or from the serosity of the blood. (§ 13, p. 48 *s*.) Some also from the mammæ; some from the little sheaths and porous papillæ of the placenta. (§ 14 *s*. p. 49 *s*.) *De Graaf* alleges that it varies according to the period of gestation, but he supposes that it proceeds from the chorion into the amnios. *Bidloo* figures several small ducts which he found with the microscope in the umbilical cord. (§ 15, p. 51.) Vieussens, in *Nov. Vas. Corp. System.*, observes that it is elicited under the form of vapour from the nerveo-lymphatic vessels of which the cutis is made up, and the lymphs again admixed are expended upon the nutrition of the fœtus. (§ 18, p. 53 *s*.) Hoffmann likewise says that it is strained through the tunics of the fœtus, at that time very thin and delicate, as through strainers, and deposited within the circumference of the amnios; but afterwards he says, that it comes from the uterus, through the lacunæ of the placenta, and the amnios. [?] (§ 19, p. 55 *s*.) According to Bartholin, the allantois is the middle coat, lying between the chorion and amnios. (§ 27, p. 65.)

241. The *chorion, amnios and placenta, together with the umbilical cord, that is to say, in one common phrase, the secundines, are the outermost instruments, neither belonging properly to the fœtus, nor to the mother; but they are means and communicants to enable a kind of new microcosm from the macrocosm, to arise in the world. Apart from them there would be no arising of the kind; nor unless all that will yield nutriment to the offspring, were included in an egg. The chorion and amnios therefore are in the place of the shell in the egg.* The manner of production is altogether different in viviparous and oviparous animals respectively, but in principles both are coincident. Whatever the mother expedites to the fœtus, is put into the egg; and when the albumen and vitellus—the white and the yelk—are consumed, the chick is formed, and breaks the shell. A similar liquor also is extracted by the fœtus in viviparous animals, and their blood and fibres. Otherwise the mother would have to sit upon her ova, and in the meantime would be of no use to society or the domestic sphere, as the member and mother of a family. This would also bring about her death; for she could hardly survive the exclusion of her fœtus; and she would be so big, that her preservation would be impossible: in a word, she would perish. But it is clear that the secundines are only instruments, and not principles. For they do not preëxist in the womb, but grow there; and afterwards are separated and thrown aside: and on the next occasion of pregnancy, new secundines again grow. They exist therefore merely for the sake of the communication between the mother and the fœtus, and for the preservation of the latter from all external injuries.

242. *The outermost coat, namely, the chorion, includes the placenta in its duplicature, through which also its abundant vessels continuous with the placenta, run; very much in the same manner as happens in the dura mater cerebri.* The great sinuses are contained in the duplicature of the dura mater; and between its two layers run its proper vessels; and beside the sinuses, the vessels that bring back the (venous) blood of the brain. The like case obtains also in other parts; and this is the reason why the chorion is double.

243. *Thus it is enabled suitably to carry around a continuous wave of the blood of the placenta in a kind of hemisphere; and thus to dispense the forces of the blood, and prevent the placenta from laboring under either an excess or a deficiency thereof.* In this way the chorion is a kind of asylum and liberal receptacle for the superfluous blood; and hereby the placenta is always conserved in the due condition; and by this means, the embryo which it encloses is not bound to draw any more therefrom than the needs of its state enjoin. For whatever of the blood has once been brought into the placenta, is thenceforth the property of the embryo, but not of the mother, from whose power it is ever after withdrawn.

244. *By virtue of its circulation through so spacious a field, this effect follows,—that the blood can the meanwhile be conveniently expurgated; that is to say, its more unclean portion can be driven out through the pores and emissary ducts, into the adjacent uterus, or into the cavity thereof. And thus it is that the placenta projects the grosser part of its blood towards the circumferences, and thus casts it out.* Such is the use to which the chorion is devoted: as may be concluded indeed from its external surface, which is uneven, scabrous, villous, and full of similar ends of arteries. On the other hand, on the internal surface, or that which looks towards the amnios, it is glabrous, smooth, and united or continuous; and hence the spring of the liquor amnii does not come from this side. Where there is any outgoing liquor, there we meet with roughnesses composed of the ends of tubes; none of which, however, are visible in this part: and indeed the amnios renders this impossible. And moreover the veins of the uterus stand with open mouths on the cavity of the uterus; and these veins are very large: when the

air is blown into one of the veins, it escapes into the uterus; and when forced in through any of the mouths, it is sent through all the veins. The application of the chorion to the parietes of the uterus also goes to prove the point; as well as also the substance of the part, in that it is spongy: which shews that this same part in the placenta, that is to say, the part that surrounds the placenta, attracts the blood from the great body of the mother; while the remaining part rejects the unsuitable portion. And by this means care is taken that no sort of flatulence shall insinuate itself with the blood or its humor, towards the infantile body; nor distend the cavity of the uterus, now that its cervix is thoroughly closed up. Nothing is more hostile than the air to the formation of the embryo, which latter sedulously guards against its approach.* If this very little world be compared with the great world, and a parallel be run between them, it appears that there is a kind of central place, a place of the utmost purity and levity, in the embryo, and its innermost parts or brains, and that its outermost region, whither all its weight and gravity tends, is that last surface which is formed by the chorion.

245. *The chorion also preserves the delicate embryonic body which it encloses, and which begins and grows from the purest substances and principles, from all possible external injuries of every description: the liquor amnii too has the like function.* These external injuries are either immediately, or proximately from the uterus, and arise from its serum, recrementitious matter, sanies, flatulent elements that exude through the pores that open into it in all directions, and which if they came in contact with the naked frame of the embryo, might soon break up its exquisitely subtile cuticular tissues; and this especially in the more unseasonable and immoderate emotions of the mother: to say nothing of very external causes, which are wont to affect the uterus from without, such as compression, concussion, constriction, falls, &c.

246. *The function of the liquor amnii, and the way by which*

* Moral and intellectual parallels run hand in hand with Swedenborg's inductions. Is not the windiness of speech carefully excluded from all that is really creative in man? The beginning of formative genius is contemporaneous with the preclusion of the surrounding flatulence.—*Tr.*

the embryo is nourished from the beginning of its formation to the end, are points that can only be known from principles: for in duly exploring matters like these, causes require to be conjoined with effects, and effects with causes. The minds of the learned are distracted and dissentient, not only respecting the source, but also respecting the use of the liquor amnii; and indeed also respecting the channel through which the embryo receives its nutritious juice. I will not however recount the multitude of opinions, for fear I should appear to pave my course by a process of destruction and refutation; which I regard as an unworthy method. To attain to the exploration of better truths, we require indeed not only experience, but this experience must be in association with principles; and conclusions may legitimately be inferred from the connexion of these two.

247. *First it must be known, as a matter of course, whence, and what, the elements and principles are from which the embryo, both to be formed, and already formed, grows and receives its nutrition: we must know that it does both from the blood alone and the spirit of the blood. We must also know what the blood contains, and what the spirit is; and how from these the fibres and vessels are formed; and all the organization of the body from the fibres and vessels. Moreover also we must know what the state of the embryo is when enclosed in the mother's womb, and the same comparatively with the state after exclusion from the matrix. Also we must be acquainted with the state of the embryonic cutis, which is in immediate contact with the liquor amnii. Besides several other matters of which we must have a previous knowledge, before we can pronounce anything with certainty respecting the subject that we have in hand.* In Parts I. and II. we treated of the blood, and shewed what it is; and in Part VII., which will be devoted to Psychology, we shall treat in like manner of the spirit. The state of the embryo is a matter which also we have discussed, and shall discuss again; as well as the nature of the cuticle or skin, and many other points which will conduce to the illustration of the present theme.

248. *The due survey of all these matters being completed, it follows that the embryo derives its nutritious juice, and the means of aggrandizing its bodily life, from no source but the spirit and blood of the mother alone. The channel of the nutrition is plainly*

conspicuous in the placenta and umbilical cord. There are in fact two arteries and one vein, to say nothing of fibres, which make up this channel. The way also is plainly visible from the umbilicus to the liver; from the liver towards the vena cava, and thence into the right side of the heart; hence from the carotids towards the cerebrum; from this again through a sort of circle towards the left side of the heart; and from this thereafter to all the viscera and over the whole compass of the embryonic frame. In the body after formation the like occurs: the nutriment is first converted into chyle; from chyle into blood; and afterwards in the blood is sent over the whole of the living bodily system; which shews that the only way of animal nutrition is by means of the blood and the spirit. And moreover it will be demonstrated, that there is nothing in the body but the fibre, that gives it any organic form; and that in the vessel also there is nothing but the fibre: moreover that the simplest fibre consists of nothing but the first animal essence or substance; for the simplest fibre is but the determination of that substance. Consequently that the fibre composed of simplest fibres is what carries the animal spirit; and the fibre composed of the second fibres again is what we call blood-vessel, and what carries the blood. Thus it is that there is nothing of formation or organization in the living body, except the simple fibre. But the primordial saline and sulphureous particles, that is to say, the first elements of the material world, do not get into the composition of the fibre, but only enter into the composition of the fluid that runs through the fibres or vessels. Consequently the purest of all the elements of the world, or the principles, enter the animal spirit, while the grosser parts, that is to say, those of a lower degree, enter the blood; thus enabling the latter to contain in its bosom as a common cabinet, whatever is required for all these humors so necessary to life; for this among other reasons, that the organic animal form may constitute a part of the ultimate or lowest world, and perform its functions suitably to these last and lowest effects of nature. Otherwise its form would be a part of the upper aura, and be an inhabitant of the heaven but not of the earth; inasmuch as its genuine form, that, namely, which is determined by the simple fibres, consequently by the spirituous essence alone, could not by any possibility play the part of an inhabitant of the earth. I have here chosen to compress these statements into a compendious form, to enable me to set forth on the ground of principles, whence and what the principles are of which the animal body is formed, and consequently by which it is reared and nourished. I have wished therefore to shew, that in regard to body, it is nourished principally from the blood, which is the cabinet of all things in its body. Apart from the blood there is no bodily life, nor of consequence any nutrition. On all these points we have already in part treated; and in part we have still to treat them. When they are demonstrated, it will follow, that there is no other possible way of nutrition than through the placenta and the umbilical cord.

249. *As then the blood is the cabinet of all things that are to exist in the body, and as the blood contains within it not only the primordial elements from whence arise all the families of salts and sulphurs, that is to say, of terrestrial substances; and as moreover it at the same same time contains the animal spirit, which embraces besides in a fixed shape the very purest principles of the world, that is to say, of nature; and in addition that prime animal essence which as regards substance is termed the soul: it follows from this, that the rudiments of the embryo conceived from the spirituous seed of the father, and arranged at once into the most perfect organic forms, resembling the innermost forms of the cortical glands of the brain, afterwards takes increment and growth from the spirit of the mother, and indeed successively first from her spirit, and next from her blood, in which latter not only her spirit, but also the other prime elemental parts are laid up.* Respecting these several points we refer the reader to our various Parts: particularly to our last or Psychological Parts. As they do not immediately concern the liquor amnii, we simply adduce them here as premises which are necessary to go before our conclusion.

250. *But when the embryo has attained its full dimensions of brain and body, and the little organs of the brain have also been rendered active, and begin to conceive and to exclude, successively more and more, the first and inmost, or the proper animal essences; then also in part the embryo begins to provide for itself, and to coöperate with the mother; receiving in fact from the mother accessory and terrestrial parts, and a constantly less-*

ening quantity of spirits. And now by little and little the embryo is left to itself and its own principles, and gradually thus struggles to escape from the prison-house of the womb. Such are the premises which we have felt it necessary to make: let us now betake ourselves with all diligence to the liquor amnii, its origin and use.

251. *From these principles it follows, that the source of the liquor amnii is by no means to be sought in the secretion of the outermost tunic or the chorion; nor yet in the amnios itself; nor yet in the bladder from the channel of the urachus. For the chorion is the outermost tunic of all, irrigated with nothing but blood-vessels; scabrous and as it were villous externally, and spongy also in this situation; but on the inside smooth and polished. Furthermore it is continuous with, and closely united to, the tunic of the amnios; in which however there are no vessels whatever to be seen, and above all, no excretory vessels, and no tubes from the deep parts of the chorion: but the passage is closed from the chorion inwards. This is in order to prevent the grosser and unclean part of the blood from getting into that interstice, and actually wounding the tender tunic of the embryo, and thus interrupting the work of formation. Nor does the urinous liquid pass in thither, for the channel is impervious in the human subject; or were it pervious, it would not lead into this cavity. In animals it passes into a particular tunic, the allantois or tunica farciminalis, which draws to it and puts away the whole of these unclean materials.* In the human subject the bladder is closed, and sometimes distended with a kind of urinous or rather serous humor; nor is the urachus pervious; or if pervious it be, it never opens into this cavity, but into the chorion: while in animals it opens into a particular tunic, the allantois, which is large, and capacious enough to hold these contents. The reason of its presence in brutes, appears to be, that brute animals arrive more quickly at maturity, both in the womb, and out of the womb; consequently more quickly attract the unclean blood and serosity of the mother; and when the blood is impregnated with this, the whole of that serous portion cannot be expelled so speedily through the pores of the chorion, but that it also passes by way of the bladder and urachus. Otherwise is the manner in which the embryo treats the humor; for the embryo

begins its woof from within, and proceeds from the first thread by comparatively slow degrees; and then for a long time it requires only the purer essences, and rejects the grosser; a circumstance that arises by the methods of the before-mentioned coöperation.

252. *And then the human embryo grows up more slowly even from the very beginning, and successively milks out the mother's juices according to the requirements of its formation, and appropriates to itself those that it milks out; nor sends back into the mother's veins through the path of the chorion any but the grosser portion. And as the infant imbibes more greedily those juices, and nevertheless they suit its blood: it follows, that agreeably to the natural circle or cycle that subsists even after formation has taken place, it expels all that portion that it does not want, towards the circumference of its body, or towards the cutis and cuticle, and this, by hosts of little arteries, namely, those oscula which exist at once with the greatest openness and in the greatest numbers on its delicate skin. Moreover, it reserves all that portion, that for the present it does not want, for future use; in order, namely, that it may provide for itself in the coming periods; and little by little accustom itself to live on its own resources, and successively as it were put on the will to assert its liberty, and to free itself from the mother, and from the captivity of the womb.* This is also a general phenomenon after birth and in adult life; for the serosities of the blood are thrown out by means of the layers of the skin. The nature of the material that oozes forth in the shape of the sweat, and also in the shape of the insensible perspiration, from every particle of the body, may be seen in the works of the anatomists, and also in our Part on the Skin. Now as the skin of the embryo is exquisitely delicate, the pores stand everywhere most open upon it; and the vessels are most abundant, so that nearly the whole expanse is ruddy with them: and moreover as the urachus is not pervious; and none but heterogeneous elements are sent away through the chorion; and still the better essences are continually brought back by the lymphatic vessels: furthermore since the infant lies in a perpetual bath; what then is more in harmony with the order of nature, as it exists plainly even in adults, than that this sweat should ooze from the circumference of this little infantine body. That all the matters

taken are for the sake of the embryo, see above. I find also that Vieussens and other writers assent to these views. Nor perhaps can it be denied, that there is also some exudation of this humor through the cord; if, according to Bidloo, oscula are there placed; but this only from the arterial capillaries, those namely that run outwardly, while the veins have an intermediate situation. In this place also the chorion is closely conjoined to the amnios. But I still desire to leave this point in doubt.

253. *As then so considerable a quantity of liquid is collected between the amnios and the embryonic body, it follows by virtue of the same law of order which is plain in adults, that when the embryo requires less of aliment from the mother's store, it demands this same liquid of its own, and thus provides for itself out of its collected resources, and by degrees begins to take its own counsel for its life.* This is a common fact in adults. Thus we not only reject sweats and serosity through the layers of the skin, but we also attract abundance of fluids; and this purely according to the changes that take place in the states of our minds and bodies. See our Part on the Skin. This state also, both in the embryo and the mother, is subject to change; and were the mother to supply the whole quantity to the fœtus now on the verge of maturity, the entire amount of her blood would scarcely suffice; for the want increases by degrees with the increments of the fœtus. Thus it draws from a kind of lake, in taking from this liquor. The change which this state undergoes, both internal and external, is a consequence of its vital motion and proper action in the womb; also of the distension of the space between the amnios and the body, and of the reaction of the uterus, the peritonæum, and the muscular substance of the abdomen of the mother: so that the fœtus now attempts to make the change, for gravid and pregnant reasons.

254. *In confirmation we have also the character of this liquor, in that it is pellucid, not aqueous but glutinous; not unlike the fluid albumen in an egg which supplies the chick with nutriment. The qualities of this humor have not only been examined chemically, but they are moreover indicated by the envelopes in which it is contained. Thus the amnios is a delicate tunic; likewise the cuticle of the fœtus: and in these tunics, no other than a*

corresponding liquor can be enclosed, according to a general rule. The character of this juice is such, that it contains the elements of the blood mixed with water particles: shewing that when the embryo puts forth the spirituous essences from its own proper laboratories, and the other essences are attracted from this liquor, the blood is produced from the two; as afterwards from the nurse's milk, which possesses similar parts of the blood, homogeneous with, and proper to, it. Thus at last the infant accustoms itself to take care of itself without the intervention of the mother, and to procure its own blood, so as to enable it to live in the state of separation and exclusion. The reader will find these particulars explained in another place. The case is especially to be compared with that of the egg; where the chick draws the finer and spirituous essences from the albumen, and lastly from the vitellus or yelk; and when these are consumed, it makes effort to come forth. The composition of the blood requires a proper spirituous essence, and moreover material elements. Both these, as its property, are now present to it: the spirituous essence from its own brain; and elements with aqueous vehicles from this environing lake. Nor is there anything to prevent the fœtus from drawing a part of the same liquor through the mouth and the œsophagus; for the mouth, if the lips be open, is also open to the same humor, which washes all around it: so also again it may pass into the veins, which are in smaller numbers in the stomach, which is still nearly closed, than on the cuticle. That our purest nutrition takes place by means of the layers of the skin, and that the food thus introduced passes immediately to the brain, may be seen in our Part on the Skin. To the reception of this food then the infant already accustoms itself in the womb. Something of the same kind occurs here as in many insects, which lay up a store for future use: bees, for example, and others; which provisioning of theirs is attributed as it were to no cause at all, being assigned to instinct. But this openly shews, that the soul is a being of the higher world; and the soul of animals, a being of a higher nature; the human soul, however, a creature of heaven itself; and that this soul it is that commands all these fates, and prescribes their conditions; and that it contains as present the whole series of formation, both the present things, and the future,—all that belong to the state of its bodily life. The fact however that it possesses this nature, cannot be its own prerogative in a universal sense; but there must be a certain universal being that perpetually flows in, whose property it is to communicate this faculty to it continually. This providence may indeed exist as properly its own, inasmuch as it receives it from a more universal being; just as thought is the property of the mind, and yet it receives the faculty of thinking from its soul. Sight is the property of the eye; yet the eye receives the faculty of seeing from the inner sensory. The case is the same with these powers as with instinct. The representation of all things that belong to the process of formation, is the property of the soul; and yet it receives that faculty from the universal, that is to say, from the divine spirit. This however will be more clearly evident, when we treat of the influx of higher beings into lower. Something of the same kind also occurs in the formation of the infant. The body of the mother is its universal world, which gives to the embryo the faculty of forming itself; and yet the formation is the embryo's and its own proper soul's. For whatever it receives or accepts universally, it claims as its own, and completely appropriates. *From these positions it follows, that the liquor amnii is genuine chyle, or that it contains the parts that genuine chyle should contain; that is to say, the veriest essences out of which the blood is to be prepared. Next after it comes the milk that is emulged from the mammæ.*

CHAPTER VIII.

THE UMBILICAL CORD.

255. HEISTER. "The umbilical cord is a structure twisted like a rope; as thick as a finger or thumb; and made up of one vein, and two arteries, the umbilical vessels; surrounded by a common, thick membrane, which however is soft and as it were spongy, on account of numerous cells filled with a limpid glutinous liquor; and is continuous with the amnios and peritonæum of the fœtus. This cord begins in the placenta, by many roots of veins and arteries; and ends in the umbilicus or navel of the fœtus. It is two, three, and sometimes four spans long." (*Comp. Anat.*, n. 244.)

256. TH. BARTHOLIN. "The four umbilical vessels . . . are covered in and enveloped by a kind of common coat or crust, . . . which not only involves all the vessels, but also distinguishes them from each other. . . . The umbilical vein is much larger than the artery, and passes through the double membrane of the peritonæum. . . . It ends in the vena cava, within the body of the fœtus, near the right auricle of the heart, and permeating the liver, bores its way through into the vena portæ; then from the gibbous portion of the liver it comes forth by a considerable orifice, and passes outwards by the umbilicus, sometimes double, and divided into two branches. . . . From the umbilicus it goes upon the breast, and from thence running sometimes obliquely, through either part, right or left, of the throat and cervix, it goes by windings to the occiput through the middle of the forehead gradually to the placenta. Sometimes . . . it goes round the neck, and then runs at once to the placenta. . . . The umbilical vein exhibits several very conspicu-

ous nodes. (*Anat.*, p. 310, 311.) The two arteries are inserted into the iliacs, and run together with the vein, &c. . . . These arteries do not join the arteries of the uterus, according to the most positive observation of Arantius. (p. 312.) I often pressed the turgid vessels with my finger, and observed that the blood could be easily driven from the vein toward the fœtus, but not to the placenta, where the nodules resisted like valves. On the other hand the blood could readily be pushed from the artery into the placenta. The same thing was manifest with ligatures, &c. . . . For the sake of this motion the venous and arterial branches are joined by anastomoses within the placenta; in order that there may be a ready return of the blood running back from the little arteries to the little veins. . . . Sometimes the little branches of the veins and arteries climb across each other in crosses both internally and externally; sometimes they are joined to each other by insertion; sometimes they come together at the sides; and sometimes they are twisted together in a spiral manner. The minutest branches of the vessels are put together to form larger branches, united in like manner, but by more blunt anastomoses, until the arteries are reduced to four branches, and the veins to two, which at last combine into trunks *sui generis*, and which emerge from the placenta. The arteries surround the veins, in part accompany them, and in part run alone. (p. 314.) Outside the umbilicus and placenta these united vessels as they advance and run in the cord, are twisted together in a remarkable manner; their circumvolution however is for the most part so arranged, that a larger cord is produced from the spiral gyration of the lesser ones, very much after the manner of our unicorn's horns which are made up of striæ; and this disposition of parts can be easily seen by holding the cord up against the light. (p. 316.) We have further to notice in the circumvolution of the cord, that in the vein there are transparent nodules and spots, from the blood appearing through the thin and pellucid tunics; this is not the case in the arteries. . . . At a span from the conjunction, a very remarkable texture is seen, and a rougher and as it were confused contortion. . . . The cord is of the thickness of a finger. . . . When dried it becomes much thinner. . . . When the child is born the cord requires to be tied, . . . and cut off an inch and a half from the ligature, . . . until the portion which is left dries and separates. The period at which this happens is different in different infants. . . . The umbilicus thus formed and thus consolidated and strengthened, is covered with a robust skin or cutis, which is capable of immense preternatural distension, when it receives the protruded intestines in umbilical hernia. (p. 316, 318.) The umbilicus or navel is pervious in some subjects. . . . When purgatives, to exert a sedative effect upon hysterical sufferings, to kill worms, &c., are placed upon the navel, it opens in these cases in an insensible manner.—After birth however these vessels degenerate within the abdomen into ligaments; the vein [into those] of the liver, and the arteries, into the lateral ligaments of the bladder." (p. 319.)

257. Schurig. Verheyen says: I particularly saw on the sides of the umbilical vein certain whitish and hardish bodies running out like a thread through the whole of the cord: these he says are like nerves. *(Embryologia Historico-Medica,* 1732, t. iii., § 10, p. 82.) Many persons agree in affirming that there are many nodes in the cord: the midwives make auguries from these: they are nothing more than the twistings and circumvolutions of the umbilical vessels. (§ 11.) *De Graaf:* It is a stupid midwives' tale, that the penis will be larger if the umbilical vessels are not tied close to the umbilicus. (§ 13, p. 86.) In a fœtus one month old the cord is scarcely longer than a finger; but about the time of birth it is about the length of a Paris ell. (§ 14, p. 87 *s.*) *Riolan:* Sometimes from the looseness of the cord the fœtus turns itself so completely, and so twists about, that a knot is tied in the middle; and when this is the case, it generally involves the atrophy and death of the fœtus, from cutting off the channel of nutrition: though indeed similar knots are observed in living fœtuses. Mauriceau relates cases of several knots of the kind. So also does Schurig himself. Sometimes the cord is too long; and sometimes it is too short. (§ 16 *s.*, p. 90 *s.*) Authors also make mention of defective umbilicus, but its vessels were supplied by the umbilical vessels inserted into the base of a kind of a tumor; where a portion of the cord was seen, bound down in the ordinary manner. (§ 18, p. 94.) A double cord has been also observed: see J. Gulielm. Riva, and other writers. (§ 19, p. 95 *s.*)

258. Those *two tunics, the chorion and amnios, appear to be the first-born tunics of the ovum itself, which also is surrounded by little tunics; but afterwards to be increased in thickness and capacity according to the increment of the fœtus they enclose.* Thus the fœtus must be said to be concealed in its egg or ovum, so long as it is within these membranes: almost in the same manner as the chick in its egg. The whole difference arises from this, that the chick possesses all its supplies of food enclosed within the egg; while the embryo or vivipara has to draw the same from the store which surrounds it, or from the breadth of the mother. Hence the chorion is a spongy and soft tunic, whereas the egg has a hard shell. The amnios corresponds to that very delicate bark, philyra or skin that we find within the shell; but there is a difference between the albumen of the egg, and the liquor amnii of the fœtus; for the albumen contains at the same time, the first essences, even the spirituous essences of the chick, while the liquor amnii contains only lower or latter essences, those namely that conduce simply to the constitution of the blood.

259. *The umbilical cord seems moreover to have preëxisted anterior to conception, and in the primitive ovum, but only to have been a sort of continuation of the inmost tunic; for which reason indeed it is still continuous with the inmost, namely, the amnios. But before conception it was only a slender little tube, and at that period destitute of its fibres, and also of its vessels, both arterial and venous.* When it lay in its first intestinal bag, namely, the corpus luteum, or was applied upon the top of it, it appears at that time to have served as a pore or passage for the juice or albumen that was contained in the ovum, and to have been con-

tinuous with the inmost tunic of the ovum; perhaps also to have served as a sort of exceedingly fine filter for the influent juice that was secreted first in the corpus luteum, and thus supplied to the ovum. On this subject however we can state nothing with certainty, nor can we do more than conjecture in some measure from analogy, and the visible sphere.

260. *Wherefore all this apparatus, or these secundines, seem primitively to have been the mother's, but not properly the embryo's; for they existed before the conception of the embryo; afterwards also they seem to have come together by the very vein from the substances of the mother.* As indeed experience testifies. The innermost coat of the uterus lies upon the placenta, and passes thence, almost in the manner of twigs or offsets, to the branches of the vein, and at length to the vein itself: see above. Which shews that primitively all the extension of the cord from its junction with the placenta to the liver of the infant, was the mother's: for this is the universal way, that is to say, by this way is determined the universal influxion that takes place from the mother. *The rest however belongs to the embryo; that is to say, both the fibres and arteries, which so increase in process of time, and take on such new forces, as at length to claim all power of acting for themselves, and wrest it from the mother. In this way that becomes the property of the embryo which at first was the property of the mother, and afterwards was common to the embryo with the mother.* See above, and also the sequel.

261. *The umbilical cord is inserted in the middle of the embryo, and as it were in its middle focus, umbilicus or navel, which stands at an equal distance between the highest and the lowest parts; that is to say, in the middle between the hypogastric and the epigastric regions; and so soon as it gains this point of insertion, it is conveyed in the first instance to the most quiet place of all, that is to say, to the bladder, and therefrom, as from a kind of centre of quiet, towards the liver.* We shewed above, in the Chapter on the Bladder, that the bladder is the most quiet place, and as it were the centre of all the abdominal viscera, which relatively to it must be said to be placed in the circumference. Hence, from the path which the cord pursues through the body of the embryo, we may come to a conclusion

as to where the axis of the whole body lies, and particularly the axis of the lower body; for the cord runs through nothing but axis. But the peculiar relation of all the other viscera to this axis, is not easily comprehended, unless we investigate the perfections of the higher forms, of which we have treated in a special doctrine.*

262. *The outermost coat of the cord which is continued from the amnios, the moment it touches the embryo in the point of the navel, is continued over the whole of its skin or cuticle, and afterwards over its peritonæum; consequently it is continuous with all those membranes that surround the fœtus: and in this way, by means of the little arteries, with the cerebrum, cerebellum, and both the medullæ or marrows; that is to say, with all the active principles of the new embryo.* See what we said above on the subject of the skin; to wit, that there is an immediate communication between the skin and the brain, and that by this channel, the animal draws in the purest elements,—elements of which the animal spirit is composed so far as relates to its bodily part.

263. *And if it communicates with the brains and marrows, it also communicates with all the fibres, or by reflexion from the brains, with the whole organic system, which is a coördination built of nothing but fibres.* Everything originates from fibres, and there is nothing whatever in the body, so far as it is animal organic and a form, but is due to the fibres and their spirit: see above.

264. *But this communication with the fibres is an oblique communication: there is also one that is more direct. For the fibre produced and brought forth from these its sources, projects itself in abundant force first of all over the whole compass of the body; and thus communicates by perpetual anastomoses with those [fibres] that pass from the embryo through the umbilical cord together with the arteries all the way to the placenta.* The reader may see in the Plates of Eustachius what a vast number of nervous fibres there are that are sent forth into the cuticular circumference of the body. In fact it is dotted over with infi-

* See the author's doctrine of forms as illustrated throughout his works, and especially in his Part on the Fibre.—*Tr.*

nite quantities; and these fibres, grouped and rolled up into glands and into vessels, constitute the very substance of the cutis. In the first period of formation, all passes directly from the first to the last, or from the inmost to the outermost sphere. The outermost however are not relatively the last of the order apart from the intermediate, but they are the next or proximate; for the cutis is exquisitely tender, particularly in the embryo; wherefore it is the next degree. It is clear from the anatomy of embryos, both in the case of the ovum and the chick, that the membranes are the first parts which are visible, and that in due order the viscera are formed within them agreeably to the relations of use. There is therefore a direct communication between the umbilicus and the fibres of the cord. Bartholin seems to think that he has observed these fibres. Moreover where there is artery, there also there is fibre; for the first and innermost coat of the arteries is fibrous, and thus the fibre is a proper part of the embryo.

265. *And if we may judge from the adult, the fibre of the cerebrum, medulla oblongata and spinalis alone, and not the fibre of the cerebellum of the embryo, appears to pass into the cord; and in fact to be all in all in the matter of these communications.* The fibre of the cerebrum and the two marrows passes especially into the cuticular circumference, and for this reason indeed it is, that the skin is so exquisitely sensitive. The cutis also it is that passes into the organs of taste, of smell, and in part also of hearing. The fibres however of the cerebellum, or of the intercostal nerve and the par vagum pass particularly into the inward viscera, which are destitute of sense. Thence it is, according to the experience of Malpighi, that the cerebrum is later in coming, or exists after the cerebellum.

266. *The cord is also continued from the arteries of the whole of the embryonic body; consequently also by this channel, from its fibres, and from its sanguineous system.* The two umbilical arteries come off from the iliacs or from the two first bifurcations of the aorta, with a view to this communication being rendered surer and larger. By this channel also the cord communicates with the cerebrum, for the last branches of the arteries end in the cortical substances of the cerebrum. See Parts I. and II.

267. *There is then nothing in the embryonic body, from the first to the last, that does not communicate with the umbilical cord, and by this means with the placenta. So that the communion is so special or singular, that no change of state can exist in the whole embryo, that is not at once communicated to the placenta. Whatever arrives at the principles of the fibres, arrives at so many minute intimate sensories; consequently every change of state of the embryonic body, is as it were rendered present, or represented, in the placenta, which is bound at once to put on a corresponding state. Yet is the sensation not such as there is in persons in the waking state after birth; but it is like the sensation of the cerebellum relatively to the viscera to which its fibres go; a sensation namely that does not reach any distinct idea of the mind, and therefore is called a natural instinct.* For while the cerebrum still lies shut, and its interstices have not been opened, nor have the lungs breathed, they are as it were in the deepest sleep, and live under the auspices of nature; and nature acts from inmost towards outer, but not from outmost to inner, or from the external senses to the internal, as in waking persons. This latter is perverted order, unless it be constantly kept in a manner passive under the auspices of the internal senses. If we act upon the inward actively, the life is rational indeed; yet is it much more imperfect than natural life; for by it natural progression and order is too commonly immensely perverted.

268. *Lastly, the cord is continued to all the abdominal viscera by means of the peritonæum.* It follows the peritonæum all the way to the bladder, and in this way clothes the bladder. The manner in which the peritonæum communicates with the viscera enclosed within it, may be seen in our previous Parts.

269. *But the communication of the embryo with the mother, is very oblique; it takes place, that is to say, through the umbilical vein with the liver; and thus by a kind of venous duct with the vena cava and the heart. In this way the liver itself appears to be in great part made over to the maternal fibre. Yet there is no direct communication on the part of the mother with the active principles, we mean, with the cortical substances of the embryo: but the communication is merely oblique and effected by contact.* This follows not only from the connexion of the parts, but also from the effects which ensue from the mother's imagination,

which requires to be excited to the very highest degree before any sensible communication between mother and embryo takes place.

270. *But this communication decreases by degrees, and at length is obliterated; and by little and little the embryo is left to itself, to its own nature, and its own principles of life and motion.* On these subjects see what we said previously. *At last it is completely obliterated; and then all force accrues from the embryo as its own proper force; and it becomes such as to be capable of being separated, and then the infant is left to itself.* This occurs several months before birth; as is plain from fœtuses of seven months, eight months, and nine months, which survive after birth, and grow to maturity. However, as for the degrees of natural diminution and separation, we can only infer them from the facts of the case, and a little from causes; for instance, in regard to the latter means of inference, we know that the fibre of the cerebellum passes into the liver; and the umbilical artery begins to dominate over the veins, and to manage all things on its own account, so that they shall be duly supplied to the placenta.

271. *After birth however all particular communication between the body and the umbilical cord, is lost, while the universal communication is still left; and now therefore the cord degenerates, and suffers consolidation into ligaments.* The part of the cord which after its division is attached to the umbilicus, drops off of itself a few days after birth; and then also perishes the communication of the artery in the cord with the artery in the body; and also of the nerves: for at this period the child regards itself alone and its own periphery of skin, with the viscera therein included; but no longer has reference to any extraneous placenta, or by means of the placenta, to the mother.

272. *As then the fibres of the whole of the embryonic body, and also the blood-vessels of the same, stand related to the umbilical cord as a middle axis; and through this aim at the placenta as their ultimate term or boundary,—it follows that the embryo, by means of the umbilical cord, disposes and manages the placenta, according to all its own changes of state: so that the placenta supplies itself with so much, and with such a species, of fluid, as the exact state of the body demands: for the placenta is only the*

expansion of the umbilical vessels; and the communication is of such a character, that the one is altogether communicated to the other. Anatomical descriptions shew that the placenta is the expansion of the vessels of the umbilical cord; for the arteries therein, as well as the veins, expand separately, and split into the finest threads. The arteries also unite with the veins in divers ways; by application, anastomosis, superegestation, insertion direct and oblique; as Thomas Bartholin reports.

273. *In this way it is so disposed as to suit every state and change of state which the embryo undergoes; and this state is communicated to the placenta by means of the fibres and arteries: shewing that the blood or nutritious juice is communicated from that store in the placenta, according to all requirement and need: and also from the universal store-house of the mother, to the placenta, in successive order.* All this follows and takes place by virtue of the connexion of parts. As every change of state is communicated to the placenta, (see above,) and the placenta draws the whole of these resources from the mother according to its general need; it follows that the connexion is of such a nature, that the embryo commands the placenta, and by this means, the mother; and thus every want is supplied. But to give all these points in full detail would extend over several sheets. For in this case we should have to explore the connexions of the several parts in the embryo, the cord, the placenta, and the uterus. The existence of a connexion and communication of the kind in the cord and placenta, is evident enough. In both, the artery communicates with the veins, but in different ways. Whatever has undergone contact, causes a most exquisite and intimate sense to be present in the principles of the infant: and according to the knowledge in this sense, the soul of the infant disposes, and takes notice also plainly how to dispose. If these particulars are duly considered, we may to some extent elucidate the reason or ratio of influences and harmonics,—namely, how they are ordered—*constabilitæ*—both between universals and their singulars, and mutually between universals themselves. But this matter is too deep to allow of being delivered thus slightly. We do not learn anything from bare exposition, but from confirmation accruing out

of facts and multiple collated experience ;* which requires not only space, but also doctrines, in this our state of ignorance.

274. *The cord is as it were the rudder and helm, and the embryo is the mariner that governs it; and by its means so manages the vessel of the placenta, as to bring it into a state corresponding to its own: and this, by a governance and captainship of various degrees, innermost, middle, and outermost.* All the communication between the embryo and the secundines, and by these means with the uterus, takes place by means of the umbilical cord, which occupies the place of a helm; the expansion belongs to the placenta; yea even the expansion of the tunics, namely, the chorion and amnios. All these are to be regarded as continuous peripheries or as it were surfaces, but the cord is the axis.

275. *The innermost or intimate communication and correspondence between the embryo, the placenta and the uterus, is instituted by means of the fibres, which are prior to, and take precedence of, the vein, and also the arteries; and through which the genuine spirit or prior essence of the blood is conveyed.* Of the fibres we have already treated; and we have shewn that they in the first instance enter into and make up this cord; and also that the spirit constitutes the purest organs of all.

276. *The middle communication or correspondence is instituted by means of the blood, both venous and arterial.* This is the next order of communication and fluid: for the arterial vessel is made up of nervous fibres, and the blood itself principally of spirit: see Part I. Hence this communication is the next. An entire series is required before the communication is perfect.

277. *The outermost or last communication or correspondence is brought about through the external governance of the umbilical cord; for this is determined in such a manner, that the embryo manages the placenta according to every general change of state, causing it to supply exactly such a quantity as the common state requires. The embryo seems to manage this helm in the general*

* These appear to the Translator to be golden words for the direction of our talents and the prosecution of the sciences. "*Non enim discimus ex nudâ expositione, sed ex confirmatione ex posterioribus et multiplici collata experientia.*"—*Tr.*

manner, not with its hand, but with its neck. The umbilical cord follows the middle of the breast, and applies itself to the neck, running at length towards the forehead or head in the placenta. Sometimes also it passes round the neck. The embryo also makes various movements, and as it were changes place. This appears to be entirely for the purpose of guiding this helm in a general manner, and in fact either drawing upon or relaxing the placenta, and causing it to supply more or less of blood, or to draw more or less from the womb. Such appears to be the reason of the vast number of changes and motions which are sensible in the embryo; also of the twisting of the cord, and of the knots that somtimes occur in it; also of the fact that it sometimes passes round the fœtal neck, almost as though it intended to strangle the infant in its coils. These circumstances govern this state of tension in a more elongate and gradual manner, enabling the embryo to manage it according to its states.

278. *This fact of management therefore appears to be the cause of the very various motions that the embryo makes in the womb during the last months of gestation; and to be the cause also of the knots which are often seen in the cord; also of the twisting of the same around the neck, and of its various determination from the placenta to the umbilicus.* This follows as a consequence from what we said just before.

279. *All this, however, cannot be accomplished without an active force, nor consequently without motion. The real active force belongs to the embryo; and takes place first through the fibres; second, through the arteries, and the pulse of the same; and third, through the reins or leading strings of the cord itself. But the principles of the activity, all and singular, are the property of the embryo itself, but not of the pregnant mother.* There is no room for doubting this, so far as regards the fibres and the transflux of spirit; for nothing exists either animal or animate except the fibre. Nor can we hesitate respecting the pulse of the arteries, which is communicated to the whole of the placenta, into which the arteries expand; for this expansion is no other than the expansion of the arteries and veins of the cord. *It is therefore the system of the embryonic body that requires more or less of juice, and such a sort of it, or such another,*

—in short, that thirsts, hungers, demands, and invites the precise juice that it wants. Thus all is in a kind of equilibrium; whence the dispensation of singulars by virtue of the bare need of the particular and universal states.

280. *In order that all these things may be carried on with a kind of spontaneous nisus, and in an order most suitable to nature, the purest essences of the blood,—nay, and the blood itself, we mean, the purer blood, does not run through in straight lines, but in a kind of perpetual circulation. The veriest form of the purer parts is perpetual circular, or spiral. Wherever therefore the fluids have the freedom of determining their path according to their form of fluxion, this takes place according to the form inherent in their parts; and the volume of these determines itself through a course of fluxion by virtue of settled laws of nature. This is a fact which we meet with everywhere throughout the entire organic body. So all resistance is taken away; and then the blood flows of its own nature.* The fact that the umbilical cord is twisted into perpetual spires, and this, naturally, is well known to all anatomists. And further, that from the very first beginning such a gyre was established; as is plain from the first intestinal body, namely, the corpus luteum, which is twisted into gyres of the kind. And the same is shewn by all and singular the parts of the body, which as it were love this circumgyration: also again by the little arteries which are visible to the microscope, and which are twirled into perpetual gyres; so that in a space of the bigness of a drop Leeuwenhoek observed more than a thousand of these little serpents. In this axis of the umbilical cord, the blood and its spirit are at liberty to institute a fluxion in accordance with their whole nature. Hence this fluxion takes the form of an everlasting spiral, agreeably to the nature of the fluxion of its parts: of which we shall speak presently. Thus it will be declared in our doctrine of order, that the most perfect of nature's entities tend into the supreme determinations or forms, such as I term the celestial forms; and the next succeeding into vortical forms; the next again into spiral; and then the next into circular; and lastly, the next into rectilinear and angular. The simple fibre is self-determined into the celestial form, the compound fibre into the vortical form, and the vessels which convey the purer blood, into the

circular form. Only the serosity that abounds in the blood naturally tends to the straight line. Wherefore they flow, in proportion to the accompanying serum, into this last angular form, which however is against the will of the blood, and only forced upon it. Therefore as soon as ever it is freed from its serosity, it flows into a spiral or perpetual circular form. Here, for instance, through the umbilical cord. This is a sign that the purest blood of the fœtus runs through the cord, and was the blood that paved this way. Thus its resistance is of the least possible amount, and its determination natural; nor is there any repercussion, such as would be present if the current ran in a straight line at the same time that the blood were attempting to pass in a circular one. Respecting the circular volution of the particles of the blood, see Leeuwenhoek and Part I.

281. *From the particulars that have been discovered by anatomical experience respecting the umbilical cord, it appears that at the earliest period the cord, as well as the skin of the embryo, exudes that liquor which we term the liquor amnii, and which is the genuine chyle of the body: and that in the subsequent period, the cord in company again with the skin, absorbs the same, and therefrom rears and nourishes its blood.* The umbilical cord is environed with a coat that is continued not only from the amnios, but also from the skin of the embryo; and thus, by means of this envelope, the body of the embryo, that is to say, its surface, communicates with the tunic of the amnios. The cord is moreover very thick and spongy, so that in like manner as the skin of the embryo it can exude and likewise absorb a liquor of the kind. It appears to be spongy on the inside, with vesicular substances intermixed, which are distended with a corresponding juice. It also swims in the liquor amnii. According to the phenomena presented on injection and compression, there appear to be pores and ducts, or lesser lacunæ, through which the communication is brought about. From this we may conclude obviously enough, that its state is very analogous to that of the cuticle, or rather is intermediate between that of the cutis and that of the amnios. If it exudes anything, or absorbs anything, an immediate communication has place with the cuticular circumference, and also with the peritonæum, of the embryo, in which a similar chyle, nay, liquor, abounds from the very earliest period.

CHAPTER IX.

THE STATE OF THE EMBRYO IN ITS INITIAL STAGES CONSIDERED ON GENERAL PRINCIPLES.

282. The *most eminent writers from the earliest periods of history until now, have discussed the question of the relative shares that the father and the mother contribute to the life of the offspring; and whether the soul is infused into the very rudiment from the first moment of conception, or whether it is afterwards put in; whether in short it comes by engrafting, or by inspiration. Inasmuch however as we have a great desire to reason and decree on the subject of the soul and the life of the body, before we have duly considered and discussed the objects which are nearest to us, and which lie in our path and intervene between us and the soul, so do we dissert on the former themes as on objects in a most profound and dark abyss, whither our eyes cannot penetrate. And the consequence is, that we have confirmations, doubtings, dissensions, hypotheses, for which our judging faculties gain favor; procured to us, and exalted, as they are, by experience, doctrines, and the endowment of reasoning; or in the absence of these, we have recourse to a sort of enlightenment from an inward light; or more truly speaking, to an overshading or appearance from the external sphere; also to authority and unintellectual belief; not to mention other grounds.*

283. *The truth however is, that before we can declare anything in a rational way respecting the propagation of the soul, as to whether it takes place by transplantation or by infusion, it is indispensable to know previously what the soul is, what the animal spirit is, what the blood is, and even what the body is; for the*

spirits are subordinated to the soul, the blood to the spirits, (the animal spirits as they are called); and the body properly to its blood. And when we comprehend what the soul is in general, and thus comprehend indistinctly, we ought then to know what its determinations are; also what those of the animal spirits are; and lastly what those of the blood are. The former are the fibres, and the latter are the vessels. Thus at length we come to enquire into the manner in which the soul establishes its intercourse with the body; and into the nature of harmony, and also of disharmony; also of influxion, regarded in its physical aspect on the side of the body, and in its metaphysical aspect on the side of the soul. Influxion does indeed involve a certain materiality; and yet all materiality is inapplicable to the soul regarded as an immaterial and a spirit. But nevertheless influxion is to be predicated in the way of analogy; for it is evident that communication and harmony do exist. But I shall fully explain these matters in my Introduction.

284. *Therefore until all these points have been explained, we shall by no means occupy ourselves with deciding where the soul of the offspring comes from; whether it be a transplantation or slip from the father or male begetter; or at the same time from the mother or the female: or whether it be divinely poured in. In order to bring out these particulars, when we have sublimated our intellect to that point that we know what the soul is, it will be indispensably requisite, that we should explore what it is that the genital members of both sexes carry and import; viz., the testicles, epididymides, vesiculæ seminales; and by the teaching of these parts and others besides, what the semen also involves: and then again what the correspondences in women are. We must advance in order from the last to the first, or from effects to principles; and when at length by following this path we have arrived at principles, we must descend from them to effects; thus we must descend from before and from above,*—a priori et ab altiori, *as the soul and spirit themselves do, which descend from causes to effects, from the first nature to the last, from the very first elements of the sciences to the rudiments of the same.* This is the reason why I am bound to run in the first instance over the field of effects and the anatomy of the body.

285. *We may not however climb immediately from effects to*

principles, from the body to the soul, and from the material world to the immaterial; and therefore for the purposes of this ascent, I have been obliged to conceive as it were from the ovum, to form and give birth to, new doctrines, to lead me from the bottom upwards, or from the lower sphere to the higher. These doctrines I shall term the doctrine of forms, the doctrine of order and degrees, the doctrine of the society of coördinates, also the doctrine of representations, and of correspondences, and lastly the doctrine of modifications. All these require to be laid down and explained, before I dare to mount, or to attempt, that ladder that leads from earth to heaven. Our Introduction to Rational Psychology, in Part V., will deliver these various doctrines. But how distant from these sublime heights, and how far below them, we, inhabitants of earth, are placed; and still we confuse ourselves with attempting to climb at once to the very highest; while meantime we meet with infinite things on the way which require to be explained and revealed before we can reach the threshold of this temple.

286. *When we shall have gone through all this, it will at length appear, that somewhat of the soul of the father produces the initiaments of the embryo, and the principles of all that will exist in the new body. Also that these principles, which reside in the soul alone, and are now its property, successively call forth and beget, by a mode of supereminent generation, their likes and fellows: all which may thus be said to be conceived and procreated by transplantation from the soul of the parent. And that with the exception of these principles, which govern and produce that universe, all the rest are the mothers: shewing that the principiates exist from the soul of the embryo and the soul of the mother conjointly.* Principles are all those things that exist in the innermost; here for example in the innermost of the cortical substances, which are the intimate sensories and motories. Nature begins from the innermost, and then passes outwards. Not so were the soul infused after the formation of the inferior or posterior spheres.

287. *With regard to the formation of the embryonic body from the ovum until birth takes place, the following particulars flow in as consequences from the laws of nature: viz., that the formation of the body advances from the first and innermost*

sphere, in order, and by degrees, to the last and outermost; consequently from the soul to the viscera of the body. Hence all things arise successively, and there is no real type of the greatest in the least, save a representative type; that is to say, the soul represents to itself as formed that which is to be formed: whence according to the order of these representations all things follow each other and come into existence in succession. Comparatively speaking, the case is not unlike that in which the mind embraces simultaneously in one view several actions and effects, from the first cause through the middle to the last, that is to say, a whole series; which parts however succeed each other in their own order, and everywhere in that order the ends remain the same, that is to say, first ends, middle, and last; nature only advancing through effects, of which there are moments and degrees. And when the soul advances thus, that is to say, from the very principles from which all natural things proceed, nothing can hinder the just effect from taking place.

288. *Thus in the first instance the intimate and eminent little organs or sensories of the brain are produced from the spirituous seed: next the fibre is produced therefrom, or the first thread of the medulla oblongata and medulla spinalis. Afterwards the cerebellum is initiated into taking a share of the works; and the cerebrum and cerebellum together, environed with a web of the most delicate kind, constitute the carina. Therefore, after this highest region of the body has been first formed, the others that are under it, and which are external and inferior to it, are formed in their own order of succession, and take their origin from the brains as their Olympian heights. Nothing in fact exists beneath the cerebrum that is not primitively shapen by the fibre of the latter: it is the fibre, that is to say, in a broad sense the cerebrum, that prevails and governs universally. After this for the first time comes the little heart, which sustains a vicarious function in the kingdom of the body for the brain and the soul: the heart that is to say rules the blood, while the brain rules the animal spirit, and the soul rules both. Thus all discord is at length turned into concord under the auspices of the soul. Whence the mutual love of all. Because whatever there is in the body is its own, because it is the soul's; which is present indeed all over, but not in one way only, but in every fresh part in a new way.* But

for the order wherein all the parts mutually succeed each other, we must rely upon the anatomy of the chick in the egg, instituting however a cautious comparison between oviparous and viviparous animals.

289. *In order that the work of formation may be carried on properly, and according to the order of nature, the embryo immediately on the emigration of the primitive ovum from the ovary, is left to itself and its own nature. For nature acts from the innermost principles, and from her own principles, that is, from the highest principles, and from the soul itself, in which these principles naturally lie: nature acts, that is to say, in the most perfect manner. Wherefore the embryo is completely exempted from the right and will of the pregnant mother; which has no other concurrence in the case, than such as is that of the superior universal with the inferior universal; the mother in fine concurs only in supplying all the stores that the embryo requires and demands. Were the concurrence other than this, the formation would soon go to wreck, for the mother is carried away not only by the principles of nature, but also by the principles of will, which latter constantly prevail throughout the day; and are formed out of the last and hindermost sphere, in short, out of the objects of the senses; which objects are fallacies, and distract and puzzle not only the animus, but also the mind; consequently pervert the blood and the spirit, and alter their circles every moment; by which violence the delicate fabric of the embryo would soon be torn to pieces.* The animus of the mother does indeed flow into the principles of the embryo, but before this can take place, the forces must be carried up to the very highest pitch: such being the conjunction of the mother's fibres with those proper to the embryo: a subject of which we spoke above.

290. *In order then that the embryo may not be destitute of its own proper quantity and quality, whatever may be the character, and whatever the circulation of the mother's blood,—in order to this, the embryo has formed the placenta, and has conjoined it with the uterus only slightly and on the surface; and made the placenta to be a receptacle and larder from which it can draw its essences, that is to say, the blood, serum and spirit: finally also the embryo provides for itself an abundance of chyle, in the existence of the liquor amnii; in order that it may draw upon its own*

stock for what it wants, and not upon the mother's. But upon these subjects see above: for whatever the fœtus derives from the mother into the placenta, in a word, into the secundines, becomes its own property, so soon as ever it passes over the threshold of the placenta.

287. *The embryo does not always imbibe pure clear juices, but sometimes comparatively impure juices from the mother's stock; and nevertheless it requires blood and spirit of exquisite quality. Hence it has formed for its own use purificatory organs; filters, that is to say, and divers strainers. The first essences, examined and purified in every way, it draws from the placenta, in which all the vessels, venous as well as arterial, are in the first instance divided to extreme subtilty, and reduced to order: afterwards it carries this blood, which has not been sufficiently refined, to the liver; in which a second time the embryo purges this blood, and examines it. Thence first it sends the blood into the vena cava, and through a remarkable valvular opening into the auricle, and through the foramen ovale into the heart's left ventricle, and therefrom to the brains, which derive the purest essences, extracted in their cortical glands, into the fibres; and the other portions, which are in great part deprived of their spirit, the embryo sends into the sinuses of the dura mater, and then through the jugular veins into the superior vena cava, and from this into the right auricle and ventricle; and thence through the trunk of the pulmonary artery and through the ductus arteriosus into the descending aorta: and on the way towards the iliacs into the thymus gland, where again it is purified and secreted. Lastly, the superabundant viler portion of the arterial blood is forwarded by the iliac arteries through the umbilical cord into the placenta, where it circulates after having been again well mixed with venous blood; and by means of the tunic of the chorion rejects the heterogeneous and unclean elements, whereby it is again reduced to homogeneity: and so forth. This circle that the blood describes is in a manner double, but reflex; and yet it is continuous and resembles a kind of spiral. There is moreover a still more simple and prime circle of the blood; namely, when the corculum is not yet united and conical, but consists only of three vesicles. And again there is a circle still more compound; that is to say, after birth, when the foramen ovale and ductus arteriosus are shut, and the lungs*

opened. Respecting these circles, which are three, and come one after the other, see Part I., On the Circulation of the Blood in the Fœtus, and on the Foramen Ovale.

288. *The refuse substances that cannot fail to be in the serum of the blood as well as in the blood itself, are rejected principally through three ways; that is to say, the grosser among them are rejected through the chorion into the cavity, or into the gaping veins of the womb, from which they are afterwards sent away by different passages. In the embryo itself also they are rejected into the bladder, which is found somewhat distended with urine at the time of birth; the passage through the urachus not being pervious, since in man the tunica farciminalis does not exist, to receive the urinous fluid, and separate it from the liquor amnii. Finally, there is a rejection into the intestines, which are found to be infarcted with a certain meconium. This meconium appears chiefly to come down from the liver and the gall-bladder; for the first bile, separated from such innocuous blood, cannot possibly be acrid as in adults, since at this period there is nothing to excite rage or anger, or to perturb the blood, unless indeed it be any hot motion kindled in the mind of the gestating mother. And moreover it seems probable that a certain portion of nutriment, or of chyle, is drawn by the mouth of the infant from the liquor amnii; with a view, as it seems, that the stomach may get accustomed to the chyle, and the intestines to a peculiar meconium, and thus may be duly inaugurated into their uses. Experience induces us to think, that this chyle, mixed with innocuous bile, forms the above meconium. But only a very mere drop of this liquor amnii seems to be thus taken, for the œsophagus is but slightly opened: were it opened wide, in less than an hour the whole of the liquor amnii would rush into it, for there would be nothing to hinder it from so doing.*

289. *But as regards the mother, she is variously affected according to the different calls of the embryo. During the first days the embryo desires none but the mother's purest essences; then next grosser essences; and at last pure genuine blood and its corresponding serum. Nay, as soon as ever she conceives in the ovum the rudiments of the embryo, the soul admirably aware and conscious of its state, at the very first least signal, brings to the door all those things that are necessary. In fact, as soon as con-*

ception occurs, the cerebellum sends away all its spirits into those fibres that go to the ovaries, the Fallopian tubes, and the uterus; and leaves, or sparingly provides for, those others, that enter into the face, the organs of the senses, the stomach, the intestines, the mesentery, and the remaining viscera. And lastly, the cerebellum prodigally bestows the spirit upon the blood and the arteries, that the blood also may supply the same: then too it goes immediately to the blood itself, and to its serum and its watery portion. Hence we find in gravid women a multitude of successive changes of state; all of which are the effects of the appetency and suction of the embryo. These changes I will simply recount; viz., pallor, a certain intimate feeling of weakness, also an intermediate debility belonging to the internal senses; the eye becomes evidently dull, the flame and beauty of the woman are put out, and the eye is consequently somewhat darkened; a cloud hangs upon the lower eyelids: hence we find languor in the actions, or loss of the vigour excited previously at every glance when venery was kindled: sometimes also swoon; for when the spirit is deficient in the remaining fibres of the cerebellum, and also at last in those of the cerebrum, not only the intellectual sphere, but also the motor sphere of the body, and particularly of those parts of it which are supplied at the same time by the fibres of the cerebrum, are plunged into deliquium. And thus from a considerably impulsive cause we have various species of nausea, vomiting, shivering, tooth-ache, rheumatism, leanness, panting, cough, too easy falling down of the legs, thighs, feet, and abdomen: to say nothing of inordinate appetite, hardness of the breasts, swelling, indolence, and other symptoms. For when the menstrua are retained, there is no other way than that towards the breasts; namely, for the serum, water and fat, devoid of spirit, from which the milk arises. Hence the breast swells as the menses cease. And moreover the soul, which lives in the principles of all actions, and views the last sphere of things, and the middle, as present in the end regarded as an effect, arranges all things for due succession in their own order; thus, from the first initiament of the embryo in the womb, or from the first thread, to the last.

THE MAMMÆ.

THE MAMMÆ OR BREASTS.

1. WINSLOW. "Their centre or middle part lies almost opposite to the bony extremity of the sixth true rib on each side. . . . In children . . . and males, . . . they are commonly no more than cutaneous tubercles or soft verrucæ, of a reddish color, and are called papillæ or nipples, each of them being surrounded by a small, fine and pretty broad circle or disk, more or less of a brownish color and uneven surface, termed areola. In females come to the age of puberty, . . . a third part is joined to the two former, namely, a convex protuberance, more or less round, about five or six fingers in breadth; the papilla and areola being situated near the middle of its convexity. This is properly termed mamma. . . . It increases with age, and is very large in women with child, and in those that give suck. . . . *The body of the mammæ* is partly glandular, and partly made up of fat. It is a glandular substance mixed with portions of the membrana adiposa, the cellular pellicles of which support a great many blood-vessels, lymphatics, and serous or lactiferous ducts, together with a multitude of small glandular molecules, which depend on the former; all of them being closely surrounded by two membranes continued from the pellicles. The inner of these two membranes, which is, in a manner, the basis of the body of the mamma, is thick and almost flat, and adheres to the pectoralis major. The second or external membrane is thinner, forms a particular integument for the body of the mamma, is more or less convex, and adheres closely to the skin. The adipose body of the mamma is a spongy cluster, and more or less interlarded with fat. It is a collection of membranous pellicles, which by the disposition of their outer sides, form a

kind of membrane in the shape of a bag, in which all the rest of the adipose body is contained. The anterior or outer portion of this bag, or that which touches the skin, is very thin; but the side next to the pectoralis major is thick. *The Lactiferous Ducts:* The glandular body contains a white mass, which is a collection of membranous ducts, narrow at their origin, broad in the middle, . . . and which contract again as they approach the papilla, near which they form a kind of circle of communication. . . . *The Areola:* The colored circle or disk . . . is formed by the skin, the inner surface of which sustains a great number of small glandular molecules, such as Morgagni calls *sebaceous glands*. . . . They are very visible all over the areola, even on the outside, where they form little flat eminences . . . scattered at different distances round the circle. These tubercles are perforated each by a small hole, through which a kind of sebaceous or cheesy matter . . . may be squeezed. Sometimes this is a serous liquor, sometimes a milky serum, and sometimes pure milk, especially in nurses. I have seen both serous and milky drops come out at the same time. This makes me think that these holes communicate with the lactiferous ducts, and that the tubercles are a kind of auxiliary nipples added to the true ones. . . . *The Nipple* is of different sizes in different ages and temperaments, and in the different conditions of females in particular. In women with child, or who give suck, it is pretty large, and generally longer than it is broad. . . . Its texture is spongy, elastic, and liable to divers changes of consistence, being sometimes firm, sometimes comparatively flaccid. It seems to be made up chiefly of ligamentary fascicles, the extremities of which form the base and apex of the nipple. These fascicles appear to be slightly folded along their whole length, and if by drawing the fibres out, their folds be obliterated, they return again as soon as the traction ceases. Between these spongy and elastic fascicles lie seven or eight particular tubes at small distances from each other, and all in the same direction. These tubes end about the base of the papilla in the irregular circle of communication of the lactiferous ducts, and at the apex, in the same number of almost imperceptible holes or orifices; and as they are closely united to the elastic fascicles, they are folded in the same manner as they.

The body of the papilla is covered by an extremely thin cutaneous production, and by the epidermis. Its outer surface is very uneven, being full of small elevations and wrinkles, among which those near the circumference of the nipple seem to have a transverse or annular arrangement, which however is much interrupted. . . . These wrinkles or folds, like so many valves, hinder the milk contained in the ducts from flowing out; but when the nipple is drawn and elongated, the tubes lose their folds, and the passage becomes straight. Besides this, when they are drawn with force, the whole body of the mamma is increased in length and contracted in breadth, and the milk is pressed into the open tubes. Thus by barely pressing the body of the breast, the milk may be forced toward the nipple, and through the tubes. *The arteries and veins* distributed through the mammæ, are ramifications of the mammary vessels, of which one set comes from the subclavians, and is named the internal mammary; the other from the axillary vessels, and these are the external mammary vessels. These vessels communicate with each other, with those near them, and with the epigastric vessels. . . . *The nerves* come chiefly from the costals, and by their means communicate with the great sympathetic nerves. . . . The use of the papillæ and areolæ in males is not certainly known. Milk has been observed in them in children of both sexes; this was the case with one of my own brothers when he was about two years of age." (*Exp. Anat., Tr. de la Poitrine*, n. 7—24.)

2. HEISTER. "The size of the breasts is various. In virgins they are usually small: in women who are with child, or have given suck, they are largest. In girls before puberty, and in old women, they are very small; and they differ in size also in different climates and countries. The time of the breasts growing full in women, is about the age of fourteen or fifteen. The time of their decrease is various; but about the fiftieth year is the most frequent period; though it happens sometimes later, sometimes sooner. The nipple of the human breast . . . is of a cavernous substance, almost like that of the penis; hence it is capable of becoming erect and turgid. . . . It consists of reticulated nervous fibres, which are acutely sensitive, and of numerous blood-vessels derived from the mammaries; of the

extremities of the lacteal tubules, and of the epidermis, which surrounds every part of it. The foraminula, or orifices of the lacteal tubes, are . . . from seven to ten in number. The areola is beset with small glands. . . . Besides the common integuments, or the epidermis, cutis, and fat, the breasts are composed of a peculiar glandular structure of a whitish color, not unlike that of the substance of the udder in quadrupeds. This forms the middle of the breast, and is surrounded by a large quantity of fat, which makes up the greater part of the organ, and to the centre of which the nipple answers. Nuck, Verheyen, and their followers, have described and figured what they call globose bodies, corpora globosa, as distinct glands; but these are not glands, but fat. . . . Amid the glandular substance, however, there are set, besides the blood-vessels, a multitude of lactiferous tubules or ducts. These arise by fine extremities from the glandular substance; they then unite in a kind of circles by anastomosis. They are large in women who give suck, and are dilated into sinuses in many places, forming a kind of cells, which hold the secreted milk; and they communicate with the arteries and veins. . . . These parts are scarcely to be distinguished at all in very young . . . or very old women. The mammary vessels arise from the subclavian, intercostal, and axillary vessels. . . . To these Wharton adds lymphatics." *(Comp. Anat.*, n. 250.)

3. Boerhaave. "The arteries of the breasts, communicating with the epigastric arteries, form very singular knots, wreaths and spirals, and at length put forth small lactiferous tubes in straight lines. . . . There is a free passage from the arteries through the lactiferous tubes to the nipple, and from the nipple by the same tubes to the arteries again. . . . The areola of the nipple is beset with sebaceous lacunæ, and sometimes with lactiferous ducts. *(Inst. Med.*, n. 688.) Pain, hardness and tension of the breasts, erection of the nipples, and transudation of serum in the evening, sometimes happen during pregnancy. All these increase more especially during the three days succeeding the birth of the child. The drying up of the milk increases the lochial discharge. . . . The color, smell, taste and qualities of the food are discernible in the mother's milk. . . . The milk differs much at different times after fluids and solids have been

taken. . . . It is best four or five hours after a meal; whereas after about twenty-four hours fasting, it is found to be salt, unpleasant, yellowish, and soon contracts a disagreeable odor." *(Ibid.*, n. 690.)

4. MORGAGNI. "Men (males) have besides the nipple the other mammary organs of secretion. *(Advers. Anat.* V., anim. 1.) Ruysch has delineated the nervous papillulæ. *(Ibid.*, anim, 3.) I have always found a peculiar glandular substance in the breasts, compacted into a single large body, whose centre pretty nearly corresponded with the nipple. It was of a white color, had a somewhat uneven surface, and was of a hardish substance, offering considerable resistance to the knife. Within this, when full of milk, certain cells as it were presented themselves, also full of milk, and which belonged to the lactiferous ducts. . . . John Riolan also described the glandular substance of the breast in the same way. . . . But Wharton regards it as the largest of the glands, . . . and as consisting of a glandular, spongy and uninterrupted parenchyma. *(Ibid.*, anim. 4.) The glandular body we have described is contained between two soft layers of fat, which envelope it nearly all round; the fat being placed underneath it as well as above it. *(Ibid.*, anim. 5.) I have noticed particular lactiferous ducts running all the way to the several tubercles of the areola, and dilated within them, so that the build of the tubercles was in great part owing to the dilatation of these ducts, and to their projecting beyond the plane of the areola. . . . At the end of the tubercle, the duct contracts into an excretory orifice, and there opens outwards." *(Ibid.*, anim. 7.) He thinks that the tubercles should be regarded as sebaceous glands, "because a substance very like sebaceous matter may be squeezed from them. *(Ibid.*, anim. 8.) The glands of the mammæ do not consist of numerous, minute and separate glands, as authors figure them, but they all form one gland. . . . The fat is not placed within this gland, but above it and beneath it." *(Index Rerum*, p. 20.)

5. NUCK. "Arteries are distributed through the mammæ in infinite ramifications. The internal vessels, running to the cavity of the chest, when they reach the midway boundary, perforate the muscles between certain of the ribs, and in the breasts they form, by a number of convolutions, the glomerules

of the conglomerate glands. (*Adenographia Curiosa*, p. 11.) When mercury is injected into one of the orifices of the nipple, it passes into the lactiferous ducts, and also into the arterial vessels continuous with these ducts. It is clear then that the ducts have no glands. These ducts are represented in the figure like the branches of a tree. (p. 12 *s*.) They are all united by mutual anastomoses. (p. 16.) I saw the fibres, large and small, firmly attached on every side to the integument of the nipples. The lesser fibres maintain a surprising order and arrangement. The series of fibres is twofold; the one set coming from the larger retiform fibres; the other from the small. The latter are perforated with orifices of various forms, and are always connected to the larger fibres." (p. 24.)

6. WHATEVER *materials there are in the blood, that is to say, in the red globules of the blood, and the surrounding but genuine serum, must also lie in the milk, in order that the latter may serve for nourishment to the tender infant, and that it may obtain therefrom the blood which circulates abundantly in its vessels: thus if any portion of material be wanting, the infant will be imperfectly nourished, inasmuch as it uses no other food but the milk.* Common experience and analysis shew plainly enough, that the milk contains all the materials which there are in the blood, including the serum. Thus it contains the fat, which forms its cream; also the serum, which separates in large quantities from the fat; nor the less all kinds of saline particles, urinous, sulphureous, nitrous, and alkaline: thus the serum turns sour, and sometimes clots; in cheese it forms a close mass, shewing that it contains a quantity of alkaline matter. But to shew from an intimate ground what the milk must have in it as embracing all the materials of the blood, it is imperatively necessary to know what the blood itself is, what saline elements enter into its globules; also what the serum is, and what saline particles especially compose the genuine serum; lastly, what the fat is, and how the fatty mass arises out of the disintegration or resolution of the blood and serum; consequently, what the more fatty part of the milk contains; what is the composition of cheese, &c.: it would in fact require a most comprehensive knowledge of chemical science, to enable us to comprehend aright this transmutation of blood into milk. All that we can say at present is, that all the prime elements of the fluids and soft parts, and the principles of every order and degree, are

involved in the milk, which so is enabled to be the parent of all things that can possibly exist in the world of the humors; wherefore it is the larder and cupboard of all things in the body. See *Economy of the Animal Kingdom,* Part I.

7. *It is therefore requisite that the fatty matters should be combined with the watery in the blood; and this is brought about by salts of different kinds, which being friendly or akin to both these orders of substances, effect the copulation between them; which however must necessarily be slight and loose, to allow the parts to be resolved back easily into their principles.* We know from chemistry that fatty or oleaginous matters will not of themselves unite with aqueous fluids, but require salts to combine them; and the saline materials are also known: now that the copula or bond between these different substances in the milk is exceedingly slight, is evident from the spontaneous separation of the milk when left in vessels, into cream of a comparatively stiff white substance, and a sourish fluid; also by the easy separation of its elements by the motion of churning into butter, as well as by heat turning it into curds of various descriptions. The lax cohesion of its parts is also discoverable by the microscope; for it is seen to consist of globules of different sizes, large and small, each annexed to each by means of particles of salts, and by the smallest points of contact; and some of the globules are observed to be lighter and finer than others, and to rise to the surface, while others are heavier, and sink to the bottom: see Leeuwenhoek. Thus the copulation is very slight indeed, which is actually necessary, in order that when it comes to pass down and be distributed to the little frame of the infant, with its soft and most tender vessels and organs of nutrition, separation may at once take place, and the milk be distinctly converted into blood.

8. *Thus the milk is the only composition and unition of all things which there are in the blood, and the blood requires to be concentrated in it, that thereafter it may soon be able to return to blood, and describe a kind of circle like what it describes in the body; where the blood is perpetually severed into its elements, and again recompounded. Thus the milk must be a kind of middle term, which regards the blood in the mother's body as the last term, and the blood of the infant as the first.* This follows from

the nature of things, for whatever is to be converted into blood, and contains in itself whatever there is in the blood, must also consist of the several parts of the blood, and so must be a middle essence between the mother's blood on the one hand, and the future blood of the child on the other.

9. *It is very evident from many facts in chemistry, that when oily and watery substances are mingled, and that, slightly, a whitish fluid similar to milk is the result; for white color is a constant result, whenever transparent particles of different kinds are united to each other without any rule or form; which shews that it is by a necessary law, that the combination of fatty with watery substances through the intermediate agency of saline particles, gives rise to a humor of a milky color.** This indeed is supported by all the facts of chemistry, for nothing is easier than to give any oily fluid the peculiar consistence of milk by admixture with essential or other salts. But as to white, it is no color, because the parts under these circumstances are not arranged on any principle of distribution: this we have shewn in our *Economy of the Animal Kingdom*, Part I., and experience confirms it, for we see that irregular fragments of glass, ice, and many other transparent substances, colored though they be, yet instantaneously in their pieces are as white as snow.

10. *To enable us then to perceive aright the composition of the milk, we must seek it from the breasts, by an investigation of their structure, that we may see how those three principal essences are led into the milk; that is to say, the fat, the urinous serum, and finally the purest matter of the blood; for the compound which we term the milk, results from these, that is to say, from the mixture of the three.* The mode in which these elements are distinctly separated from the blood, and how they enter the machinery of the breast by distinct ways, and at last run together there, and produce the milk, cannot be better cleared up than by a full examination of the breasts. But we must not enter further at present into the chemical theory, or shew the source and qualities of the fat, of the serum, or of the intimate essence and matter of the blood: this would demand an entire course of chemistry.

* See the author's *Chemical Specimens*, translated by Strutt, p. 129—131. —*Tr.*

11. *The fact that a large supply of fat runs to the active centre of the humors, is evident from the fatty sheets and layers that surround what is called the glandular or oval body and centre of the gland; for the fat hems it in on all sides, and the adipose matter of the part is continued to the interiors, where the lactiferous ducts arise.* From the fatty covering, and also from the continuation of the fat into that middle or enclosed space, it is impossible not to infer, that this is indeed the source of the fat which appears in such abundance in the milk, and separates from it spontaneously. For the cellular tissue, which contains the fluid and yellow fat, is continued in pellicles and fine ducts all the way to its centres; and as this fat is comparatively pure, and in a manner the prior essence of the fat, and of a nobler quality than the adipose fat, so it must stream to its place through finer channels, and along an uninterrupted series of pellicles and vesicles. The common fat, however, or that which constitutes the adipose membrane, is too gross to be thoroughly serviceable for this work, and therefore a more purified fat is drawn to the spot by the above ways and passages. This is the reason why this body in the middle of the layers of fat, is at one time said to be glandular, composed of vast numbers of clustered glands, while at another time it is regarded as a single great gland. It is indeed well known that the fine fatty humor permeates purer ducts and vesicles than the ordinary aqueous humor: and Winslow confirms the view that the fat is continued to the interiors, as we have observed.

12. *But the serous part of the milk, together with the interfluent saline, acid, alkaline, and urinous particles, comes immediately from both the internal and external arteries; and as these arise from the subclavian and axillary vessels, the blood which has just come from the heart is immediately brought hither, before it reaches the emulgents and their general emonctory organs, the kidneys; and rejecting its vilest portions on the way as the arteries wind and turn, it arrives in the breast with the essence of the fresh and nearly crude chyle; this being the reason of a number of phenomena which are related of the milk of nursing mothers, and of the changes which it undergoes.* The blood as it comes freshly from the heart is filled with an infinity of alimentitious particles, many of them comparatively worthless and impure, of

which the system is disburdened directly below the diaphragm by means of the emulgent vessels and the kidneys. The subclavian and axillary arteries, however, are above this effect, and by anticipation intercept a quantity of even this impure blood, which accounts for the diminution of both the urine and the lochia during suckling, as compared with those discharges in the same persons, when the milk ceases to be drawn. This too is the reason that the milk occasionally has the smell and taste of the fluids and solids which have been lately taken as food; and that better milk is obtained three or four hours after a meal than at any other period either before or subsequently; and that it becomes too sour to be good after the mother has fasted for several hours; also, as Bartholin narrates, that a rush of milk is sensibly perceived by nurses immediately consequent upon eating: not to mention several other facts all having the same bearing. Especially noteworthy it is, that the daily practice of nursing, causes a livid appearance in the arteries close to the breasts, which seem to be turgid not so much with blood as with serum; for as we have shewn in our *Economy of the Animal Kingdom*, Part I., there is nothing more plain from anatomical investigation, than the fact of arteries laying hold of the serum, and keeping back the blood. But in order to prevent the thoroughly impure and defiled portion of the blood from passing by derivation into the essence of the milk, both the external and internal mammary arteries wind about through many and mazy turns, whereby they discuss the fouler matters; the internal mammary vessel passing in the first instance through the pectoral muscle, and purified thereby of these unprofitable substances, it advances to its destination with none but genuine and suitable serum. That such elements as we have mentioned are contained in the milk, is a point well known even to the vulgar, from the quantity of acidescent whey which remains after milk has creamed. And that it also contains alkaline, saline and urinous elements, is plain from its formation or coagulation into all kinds of curd and cheese; these being exceedingly solid and tough; and in fact what remains, and forms the last species of cheese, has a bitter taste; and does not dissolve by the application of heat, but gradually runs into a soft semi-fluid mass.

12. *From the thorough commixtion of these two essences or humors, there results an essence which is of a milky color, but which nevertheless is not yet genuine milk, until the purer material, comprising the prime saline elements or principles, which lie deep in the bosom of the blood, is sprinkled upon the compound; which is done not only in the adipose follicles or cells, but also in those follicles or cells which are seen under the appearance of glands or glandular clusters, in the middle of the substance of the breast; and also more rapidly in the lactiferous ducts, and in the papillæ.* It appears from the anatomy of the breasts, that innumerable capillary vessels beset the cells containing the fat, and according to the degree of purity in the vessels, the juice from the blood which they sprinkle upon the cells is noble and inward in its nature: it also seems, on the shewing of Nuck and Boerhaave, that these vessels make wonderful contorsions and spires in the breasts, and form themselves into knots, from which none but the purer essence of the blood can transpire or expire; thus the third essence of the blood is poured on by them; an essence which is comparatively pure, and nevertheless material. It is clear from the composition of the blood (see Part I.), that the blood-globule is made up of the purest saline elements and principles, which in fact enter its pellucid globules, and strengthen the build and frame of each; also that the saline elements of the second order or degree, in other terms are the volatile urinous and sulphureous salts, which in fact combine and fit into one globule those primitive globules of the blood (see Part I.); also that in the middle of the blood-globule there is a saline corpuscule having six concave sides (see again Part I.) This then is the constitution of the blood in its various degrees. Now for supplying the blood with these constituents, and thus combining its elements into the form of a globule, a watery serum is necessary, full of similar elements, and serving as their vehicle (see Part I.) From what we have said above then it clearly follows, that all those particulars which there really are in the blood, and in its especial vehicle, the serum, are in the milk as well. Thus the harder and grosser kind of fat is no other than the matter consisting of the saline parts or elements of the second class; no other than those urinous and sulphureous matters which fit so closely together to form the

globules, keeping however on their surfaces, while the minute volume of ether proper to each, occupies the central cavity; as may be shewn by endless proofs borrowed from chemistry. But as this fat also has saline and alkaline elements of the grosser sort inserted in it here and there, it is on this account purified in the way as noticed above, before being conveyed into the blood. But the purest principles, which are freshly poured in as dew by the arterial capillaries, form a species of primitive oil, which is as a crown to the mixture, and perfects the essence of the milk. All these substances being loosely combined, are again resolved into their purest parts, and change into large supplies of blood, as observed above.

13. *The spirituous essence, however, of the blood, is carefully prevented from passing off into this milky substance, and this prevention is effected by the arteries making innumerable turns and twists and little knots; and also by the presence of an infinity of veins, with their mouths opening wide in the cells of the fat, and also in the other follicles and glands, and in the lactiferous ducts, whereby any spirituous essence that is excreted in company with the milk, is at once most greedily reabsorbed or taken in again.* There are also an abundance of lymphatics in other parts of the breasts: and wherever they find any essence of the kind extravasated, they, as well as the veins, being present universally in the minutest pellicles and follicles, and indeed in the very nipples coral-red with blood, and in the areola surrounding the nipples, diligently seek such essence out, and seize upon it greedily, for nothing is more kindred to the blood than it; so much so, that it runs back when the blood invites it, almost as it were of its own accord, with something comparable to magnetic power. For were the milk to be full of the animal spirits, the infant drawing a few draughts would soon cause the nurse to faint away. But the infant has really no occasion to draw or borrow from its nurse her spirits which do not belong to it, inasmuch as it has a brain of its own, which conceives, prepares, and sends forth plenty of spirits; and the infant lives from the ground of its own, but not of its nurse's life. In the milk therefore there is nothing more than the material part of the mother's blood.

14. *The vessels, cells and ducts are so formed, that whatever there may be in one, is common to the others, and to all; whereby*

each part draws its supply from the common lake or stream; thus the essence of the milk can never fail to be of similar or uniform quality and preparation in all parts. This is very plain from the lactiferous ducts constantly communicating with each other by perpetual anastomoses: this is likewise the case with the arteries and veins, whose numerous writhings and knots we see; likewise with the adipose cells or cellulæ adiposæ; with the ducts which penetrate into this cavity [of the cell]; and about the nipple also there is a circle to which all the lactiferous ducts run. The same principle is of universal application throughout anatomy; and is noticed in every viscus of the body, and especially in the brain; and the object is to prevent the effect from ever failing the cause.

15. *But as nature in all her movements, in all her progress, respects an end, intending it for a first and a last, so also does she in the breasts; for she respects the lactation and thus the fresh life of the infant: hence she produces the nipple at once before the existence of the breasts, for we find it even in the new-born babe: in course of time, however, she draws out towards the nipple nearly all the arteries and veins of the breast, the fibres, and moreover the lactiferous ducts, and combines them all in the teat: thus the whole mass of the breast stands related to the nipple as the goal or boundary to which it tends.* The correctness of this proposition is proved by anatomy. For there are the same number of ducts put forth through the nipple as there are lactiferous ducts in the interior. Each of these ducts, greatly contracted as it is, regards not only its own duct in the breast, but all the ducts there, agreeably to the principle just laid down: thus the disposition is particular and general at the same time. The way in which the blood-vessels pervade this part is shewn not only by its redness, but also by actual dissection: and fibres invest the nipples in a retiform manner, in two sets: see Nuck's *Adenologia* and Plates.

16. *So that when the nipple is touched, and fondled by the fingers or lips, it is also drawn out by the suckling, whereby there is a most present communication with all the internal parts constituting the breast; that is to say, not only with the nerves which bind the intimate substances and the vessels, but also with the arteries and veins themselves, consequently with the lactiferous*

ducts, which are firmly seated upon the outspread fibres and vessels and their membranes; and the consequence is the erection of the papilla; its proneness to pour out the milk; and the incentive to prepare and attract forth new substance again for the milk. These are the ends for which the milk is prepared in the breasts. For when the external fibrous* surface, which is very sensitive, is touched, the touch and titillation is diffused at once over all the fibres of the entire breast, and therefore immediately excites the membranes and vessels, and erects the nipple. And when the arterial vessels are drawn out, and recede from the fibres, they are then in a state of elongation, and at this time they let their serum drip from them, and pour it into the milk, and similarly extend all the lactiferous ducts. And thus perforce the milk rushes out of its own accord.

17. *These particulars follow inevitably from the communication of the fibres and vessels of the nipple with the intimate substance of the breast. As soon, however, as by undue expansion and frequent use the above communication is lessened, and the firmness of the breast is exchanged for a loose and flabby character, then indeed the nipple turns yellow; or instead of being red, it becomes livid and black; and the breasts themselves fall down pendulous, very different indeed from what they are in the virgin.* The very substance of the nipple consists of fibres and vessels, and of membranes therefrom composed; hence if their connexion with the interiors of the breast be relaxed, the necessary consequence is, that the parts which were before erect and firm will collapse; and the blood collected in the vessels of the same part will not so readily find its way back; add to which, that in advanced age, for example, in elderly women, the close connexion perishes, as the use it served and the necessity for it cease; the blood, including the serum, is derived to other organs, and the lactiferous ducts collapse; and together with them the vessels and the fibres: furthermore, as age creeps on, the minutest stamina or threads are by degrees obliterated, and their sides grow together, and only the larger channels are left. Hence in old subjects, in fact about the age of 50, the breasts usually get smaller, and tend to disappear.

* By the fibres, when the term is not qualified, Swedenborg generally means the fibres of the nerves.—*Tr.*

18. *There is a peculiar connexion between the breasts and especially the nipples, and the genital members; by the intervention, however, of nerves; but which connexion has not hitherto been explored by anatomists.* This is plain from fact; for when the nipple is titillated, it not only undergoes erection for the reasons already adduced, but the same effect follows from the secret stimulus of venery, as we all know well, and as may be shewn indeed from various proofs. All that is known upon this subject anatomically is, that the mammary or costal nerves mingle with the sympathetic, or with the par vagum and intercostal; and that the latter nerves are sent to the members devoted to generation. But it is quite necessary to trace the precise communication; to be able to follow the very fibres after their anastomoses with other fibres; and diligently to mark their particular terminations or insertions.

19. *The red areola which surrounds the nipple also performs a peculiar use, for the blood conveyed to the nipple by all the branches requires to circulate round about, to enable it rightly to perform its office; and this especially during the period of lactation, when the nipple undergoes alternate extension and relaxation.* Were it not for this field and gyre, when the nipple is drawn considerably out and again relaxed, the large supply of blood in the nipple would readily rush out of its vessels, and so give rise to sanguineous serum, for one vessel is combined with another; and drawn out, extended, and in fact firmly bound up with its neighbors; thus in the meantime a gyre of circulation requires to be left to it, whereby it may have the benefit of the asylum principle; in short, of receptacles or reservoirs. Thus every contingency is met. But were it not for the areola, the communication between the vessels of the nipple and those of the breast itself would be destroyed after a few times sucking, and so the very ends of the breasts would be defeated.

20. *Yet its sphere of circulation is so limited as only to comprehend the tract where the lactiferous vessels terminate in the gland; for besides their concentration in the nipple, they also have their ends and boundaries around it, that is to say, in certain glands; whence it follows of mechanical necessity, that in no case are the lactiferous ducts dragged out of their places, but maintain their own suitable relation, connexion and order among*

themselves. This could not be the case if all these ducts, thus concentrated, terminated only in the nipple. If, however, their terminations regard also other ends or parts extraneous to the nipple, then they will not be dragged out of their relative position or from their mutual connexions, which is a purely mechanical consequence: just as when threads are combined in one bundle, and not attached one by one to any fixed boundaries in other parts, in which case they are easily shaken out of their mutual connexions. Now that these glands are the extremities of the ducts is boldly declared by Morgagni, and Winslow and others entertain the same opinion of them. For almost the same holds of these ducts as of the blood-vessels in the areola. Thus in order that the blood may ebb and flow distinctly into the lactiferous ducts, the nipple is compressed and drawn out, and by this way the blood passes back into these ducts, and thus makes a communication with the whole sanguineous texture of the breast. Nay, they excite the lactiferous ducts one by one, by a unanimous drawing or traction, to let out the milk in drops. Thus these glands are auxiliary forces to the nipple, aiding the due performance of its several offices. At one time, therefore, they pour out fatty matter or sevum, at another time serum, at another time milk, pure or mixed; and they are pierced by little openings or foramina, for were it not for these, which make their exudations possible, they would not subsist. But they at once manifest their communication, which is maintained in its integrity by fistular passages, each having an outlet. Otherwise they would be blocked up, and their cavities lost, in which case the passage would be entirely obliterated.

21. *With respect, however, to the expansion and contraction, to the explication and replication, of the nipple, and to the fact that it is destitute of valves, but its tunics, which it sets against the milk as it escapes, are wrinkled or rugose; also that the lactiferous ducts so communicate with the arteries, that no more milk is shed forth than the requirement at the end, or the force actually applied, demands; and that the superfluous portion runs back; all this is a mere mechanical or organic provision; and it is therefore evident to the sight on anatomical inspection; and for this reason we do not dwell longer upon it.*

22. *Furthermore it is to be observed, that the breast of the*

male sex also is furnished and adorned with mammæ, nipples and ureolæ, not for the purpose of giving suck to infants, but from the natural necessity of shielding, as with a coat of the thickest mail, these two sides of the chest, on account of the subjacent pectoral muscle, and at the same time that part of the lungs which lies between them, from cold and all other contingencies: likewise by a fatty expansion and protuberance of the kind, to give the muscle, and by connexion therewith, the arms, an easy power of acting: whereas, when they are returned to their places, and shaken or moved in all directions, unless these parts of the chest readily yield and relax, these actions so necessary to the human race could with great difficulty be performed. The fat is naturally carried in quantities to this situation; for when the muscles of the chest are acting, or even those of the arms, the pectoral muscle as well always coöperates; so that all the muscles of the chest make for these spots; which accounts for the quantity of liquid collected within the cellular tunic, which is the common tunic of that muscle, and also the basis of both breasts. For this reason an abundance of fat cannot fail to be concentrated in this situation. And if there be fat, and a frequent extension and relaxation of this pectoral part be required, it follows that this fat will circulate; that is to say, will be gathered near the cellular texture of the muscle, as well as round about under the skin or epidermis. And if there be this circulation of the fat, it follows that they also conspire with the blood and the ducts, or that the organs secreting the milk will stand out from the surface, and will be fixed and brought to an end in the nipple as an ultimate boundary. Conceive that in this place there were no fat and no projecting globe, but the skin tightly drawn over it as in other situations; and then conceive that the arms have to be carried back, that the subject of this imaginary formation is to toil, to embrace, to wrestle, to fight, surely the whole of this membrane would resist, or at last would be so drawn out, that when the arms were brought back, a great furrow, or protuberance, would appear. Moreover wherever motion is considerable, the facility for its execution is gained by a supply of fat round about; as in this case to the pectoral muscle, which also covers that noble part of the cavity of the chest where the lungs lie; round about are the bones, as the sternum and the vertebræ. This part is intermediate. Thus heat or cold would easily penetrate, and would trouble the mechanism and motion of the very heart itself. However that they are also formed for the sake of the milk, but as a matter of natural necessity, appears from the case of the males of other animals, which have the mammæ not under the breast, but under the belly. Writers indeed have brought together a number of cases, in which even the male breasts have given forth milk; see for many of these T. Bartholin's *Anatome* (p. 333, 334), where he cites Bodinus, Joachim Camerarius, P. Castellius, Al. Benedictus, and Ch. A. Vega, Aristotle, Matthiolus, Abensina, C. Schenckius, Jno. Rhodius, [Santorellus,] Walæus, A. Benedictus again, Nicolaus, Gemma, Vesalius, M. Donatus, Aquapendens, Cardanus; and Bartholin relates that similar cases are observed at the present day.*

23. *Now these propositions are proved by the swelling of the breasts in pregnant women several months before delivery, and particularly during the few days immediately preceding it; for the child which is now fast coming to its maturity no longer emulges from the mother through the umbilical vessels any great quantity of serous blood; but feeds through the pores of the skin, from its own liquor amnii; consequently the maternal serosity is determined in another direction; and in short, through the nearest way by the subclavians and axillaries to the breasts. Therefore in order to prevent what takes place in males, namely, the discharge of all this supply of serum by the kidneys or other channels, women have the menstrual flux. This portion, in its finer half, thus enters the milk; wherefore the breasts also come together with the menses.* But as much is here said which requires distinct experimental confirmations, we only touch upon these propositions, for the sake of putting them upon record.

* In the edition of T. Bartholin to which we have referred, 8vo., Leyden, 1673, we find that the above writers are said to record cases of milk existing in virgins, men, women not pregnant, also he-goats; but Bartholin appears to make no mention of the "cases observed at the present day," although he speaks of instances of milk existing in males "in the new world." We all know how common it is to find a milky humor oozing from the nipples of new-born infants.—*Tr.*

THE PERIOSTEUM.

THE PERIOSTEUM.

1. Winslow. "The fresh bones of the human body in the natural state, are for the most part covered exteriorly by a membrane, called by the general name of periosteum, which is extended over the cartilages and ligaments, as well as over the bones; but where it covers the cartilages, it is termed perichondrium, and where it covers the ligaments, peridesmium. . . . The periosteum in general is a fine, strong membrane or membranous expansion, not equally thick in all its parts, more or less transparent, of a very close texture, not easily yielding, extremely sensitive, and composed of several particular planes of fibres differently disposed, and mixed with a great number of small vessels and nervous filaments. This membrane does not immediately surround those portions of bones which are covered by cartilage, nor those in which the ligaments and tendons are inserted. Neither does it cover those portions of cartilages which are exposed to friction, as in the moveable articulations, channels, &c. Lastly, it does not cover those portions of the teeth which lie out of the sockets and gums. The innermost plane of the fibrous texture of the periosteum, or that which immediately adheres to the surface of the bones, is fixed thereto by innumerable little fibrous extremities detached from all the planes, and which enter the pores of the bones. These extremities are accompanied by capillary vessels and nervous filaments, which run for some way between the different planes of the periosteum, and perforate the innermost plane at the orifices of the pores of the bones. The periosteum differs in thickness in different parts; but this is scarcely visible on the outer surface; the inner however is seen to be countermoulded in many

places in the sulci, depressions, lines and inequalities on the surface of the bones. Some anatomists have been of opinion that this membrane was not only united but closely braced round the bones, to set bounds to their growth. It is probable they had only examined a few bones by this hypothesis; for had they considered those which have concave surfaces, depressions and inequalities, they would have found only a simple adhesion of the periosteum without any bracing or tension. In places where it is only fixed to the bones by the filaments of its innermost plane, the periosteum is easily detached, but this separation is more difficult where the fibres of the other planes likewise penetrate the bone, especially when these planes are numerous; and likewise where the insertions of tendons or ligaments traverse these fibres. *(Exp. Anat., Tr. des Os Frais,* n. 59—64.) That part which invests the skull is termed pericranium, and that which covers the bones of the face, or of the two jaws, is called simply periosteum. *(Ibid.,* n. 357.) The pericranium is made up of two laminæ closely united together. The internal lamina, which some have taken for a particular periosteum, covers immediately all the bony parts of this region; and the external lamina has been looked upon as a membrane distinct from the internal, and named pericranium particularly. The external lamina of the pericranium parts from the internal, about the semicircular or semioval plane of the lateral region of the cranium, . . . and becomes a very strong aponeurotic or ligamentary tent, which covers the temporal muscle, and is afterwards fixed in the external angular apophysis of the os frontis, in the posterior edge of the superior apophysis of the os malæ, and in the superior edge of all the zygomatic arch, as far as the root or basis of the mastoid apophysis. Between the two laminæ, as they separate at this place, is inserted a large portion of the temporal muscle. . . . The rest of their interval, which does not give insertion to the muscle, is filled by a cellular and fatty substance between the lower portion of the muscle and zygomatic arch. At this place, the aponeurotic hood is seen to join the external lamina of the pericranium, and they both communicate with particular aponeurotic expansions of the neighboring muscles, the sterno-mastoid, masseter, zygomaticus, &c." *(Ibid., Tr. de la Teste,* n. 199—201.)

2. Heister. "The periosteum is a nervous, vascular membrane, acutely sensitive; immediately surrounding the internal and external surfaces of all the bones, excepting only so much of the teeth as stands above the gums, and those places in the bones where the muscles are inserted. . . . This membrane constitutes the rudiments of the bones in the fœtus: it is the organ of secretion for the bony matter, as the adipose membrane is for the fat; and the bones receive from it deposit, increase and nutrition. Its blood-vessels penetrate the bones in innumerable places; as is best seen in the bones of children. When it is wounded, exostoses, tophi and caries arise in the part. Its thickness varies in different parts; but the internal layer is vastly thinner than the external, and nourishes the bones. It receives blood-vessels and nerves through certain canals in the substance of the bones by which it communicates with the marrow." (*Comp. Anat.*, n. 186, 187.)

INDUCTION; TREATING ALSO OF THE SUCCESSIVE FORMATION OF THE PARTS IN THE LIVING BODY.

3. FIRST *of all is delineated the cortical substance, as the primitive and purest brain; these cortical substances are conglomerated together, forming the first representation of the brain; it is the form and determination of this conglomeration that produces the various parts named cerebrum, cerebellum, medulla oblongata and spinal marrow.* These parts however arise successively, one before another; the cerebrum precedes the medulla oblongata and spinal cord; the cerebellum too comes after the cerebrum, as may be seen in Malpighi's treatises *on the Chick in the Egg. Afterwards, from these beginnings, fibres are put forth as appendages, but which do not yet venture to proceed to any distance. Fibres are the constituents that form the whole animal microcosm or* corpusculum; *thus the cerebrum by means of the fibres flows into and influences the entire system; there being nothing in the body beyond the fibre and its spirit, or nothing present except the cerebrum by means of the fibres.* These positions may be seen in the Part on the Fibre. In the sequel it will also be apparent that there is no continuity in the body save that of the fibre of the cerebrum. *But this fibre is variously wound in order to form the animal organism; first however, before it begins the formation, it forms the vessels of the body, to be its associates in forming the organs of the body.* The fibre apart from the blood-vessel, forms nothing that can be called purely corporeal or bodily; for blood is necessary, and vessels also, to the existence of the body; the blood however has to be prepared from the spirits or essence of the fibres with the assistance of elements brought from the circumambient world; and the vessels must be engendered by the fibres, through the winding or circumvo-

lution of the latter into tubular canals. In order then to complete this process, and to carry it back to the cerebrum, and complete the communication, there is required a bodily or corporeal fibre, to constitute the innermost coat of the arteries, and the outermost coat of the veins. But on this subject see the Part on the Skin. *By the blood-vessels thus prepared the first membranes are formed; and these mark out the primitive rudiments everywhere, and give the outlines of the tissue, but through a succession of parts.* This is seen in embryos and infants, in whose bodies all the membranes are manifestly red and full of blood; as for example in the skin, and in the membranes universally. *But in process of time these membranes cease to be sanguineous and become filamentary, or as it were ligamentous and tendinous; which occurs when these filaments are no longer traversed by blood, but by serosity secreted from the blood; in which case the membranes grow pallid.* This is very conspicuous in some vessels of large size, as in the umbilical vessels; also in those of the skin; to say nothing of the ductus arteriosus. Those vessels in fact that in the earliest stages carried blood, afterwards grow pallid and become tendinous, or consist of threads of a different kind. The extreme redness of the skin in embryos and new-born infants, nay of all their membranes, is a circumstance resting sufficiently on ocular proofs. An indication of the same species of formation remains even in adults, for the palest skin will often blush in a moment with the most intense redness, shewing that the blood is continued into threads that have been already previously deserted by it. *After these membranes, which delineate the first rudiments of the bodily organs, come the muscles, which exist between duplicatures of the membranes; but before muscular fibres can be formed, vessels of some size, arterial as well as venous, and also some bundle of nervous fibres, require to be determined to the formative membrane in question.* The preëxistence of the membrane that serves as a sheath to the muscle inserted into it, is plain from the formation of the several muscles, all of which are subsequently inserted into the prepared sheath, and there increase; as we see in the face, over the cranium and elsewhere; for the motive fibres of the muscles are made up conjointly of blood-vessels and of the genuine fibres of the brain; as will be clearly seen in our genesis of the muscles. *These muscular fibres, the blood deserting their vessels, produce the tendinous fibres, which are not active like the muscular fibres, but passive, corresponding to the activity of the former.* It is plain enough from the laws of formation (see Part I.), that the tendinous substance, which is continuous with and adherent to the muscular fibres, was itself muscular once. *From the tendinous fibres arise the cartilages, and from the cartilages the bones, soon or late, according to the order and wants of nature.* That such is the transformation, is evident from all the particulars, for the tendinous changes into the cartilaginous, and this, into the osseous, as witness only the formation of the embryonic frame, comparing it with the growing and adult body; a formation which will be plainly exhibited in the sequel. *For in order to the existence of the viscera or soft parts, minute glands must be formed, that is to say, parts as it were muscular, of the smallest size, of the fibres of the brain and the vessels, together with muscular coverings of a more general character; these glands in their turn give origin to their fibres, which are the proper corporeal fibres;* and have the task, in common with the vessels and genuine fibres, of weaving and shaping the viscus particular; as may be seen from the viscera, of which we have already treated severally. Respecting the glands that produce the corporeal fibres, see our Chapters on the Skin. We have now to advance from all these parts to the periostea. *But it is especially to be observed at the outset, that all this formation goes forward exactly according to the order of nature, and in fact successively from the most perfect form, through forms less perfect, to the last or most imperfect form. The first form of all is spiritual, and is a proper attribute of the soul; from this there is produced the celestial form, such as belongs to its purest organic substance, and in which the soul transacts the functions of a rational mind; this again produces the vortical form, or that of the organic substance, which is properly vortical in its designation, and the fibres produced therefrom aspire to take on the same form. The latter then produces the spiral form, which is that of the whole brain, also of the viscera, and of the nervous fibre in general; also of the motive fibre. Lastly, this produces the terrestrial or angular form, such as exists in the tendons, ligaments, cartilages and bones.* But on these subjects consult our Doctrine of Forms.

4. *We are now in a position to understand the origin or genesis of the periostea. For the last coat or tunic of the muscles is continued all the way to the fixed boundaries of the muscle in the solid substances; the muscle, which is a soft substance, being unable to found its forces upon any soft extremities, but requiring solids, to offer the proper resistance: hence the outermost layer, which constitutes the periosteum, is that vascular plane continued that surrounds the muscular flesh.* This membrane is the envelope or sheath of the muscle, and prior to the muscle; and thus by its continuation here it marks out the form of the parts which it is about to form. This is evidently the membrane; as we see in the loins, the feet, arms, vertebræ and cranium; for it precedes the parts that it is to sheathe, and consists of nothing but vessels; and afterwards encloses in its duplicature the muscles, or the bones, either, or both at once. And it is the same membrane that in the beginning of formation contains the pericranium and dura mater, but afterwards it is separated into two by the intervention of the cranium. And being vascular at first, in process of time it becomes white and pale, and none but the larger vessels are left in it. *The particular continuity of this membrane, [its continuations and reflexions,] are declared to us by the panniculus carnosus, and the rest of the muscles, which environ the whole circumference of the body; it is also seen in the head, in the cranium itself, even within the cranium, for the dura mater is the internal periosteum of the cranium.* From the muscular teguments it goes over the bones, attaching itself thereto, and passing about in various directions. This membrane therefore exceeds in extension that of any other membrane in the body; nay, is greater than that of the skin itself, for it is, so to speak, the first line that projects the whole web or tissue into a plane. *But the second or inner layer of the periosteum (for it consists of two membranes) consists simply of tendinous filaments drawn out from the muscular flesh: it immediately surrounds the bones, inserting and implanting its roots in fixed spots therein; thus enabling the bones to yield to the muscular force, and to submit to nature's decrees.* For all the tendons of

the muscles, excepting in the heart and viscera otherwise soft and hollow, have their tendons fixed in cartilages and bones as solid termini, with a view of enabling them to put forth their force, and to call those parts or appendages into motion: on this account indeed they demand and claim for themselves the innermost place, or that next the bone. *This layer however is not uniform or similar throughout, being in some places thick or dense, in others thin, while in others again there is no trace of it, but the naked bone begins to appear; for instance, on the temples, at the sides of the cranium, &c., &c.: this layer consists of tendinous threads set free, and as it were of a loose aponeurosis, intermixed with which we find blood-vessels, and also fibres, which are on their way to penetrate into the interior of the bones.* This is clear at once; for the compact tendons are of a hardish texture, and through them this vaginal tunic of the muscles penetrates the vessels between the layers of the periosteum, but in an attenuated form, and under the name of the peridesmium: it then splits into threads on the spot, and goes in a soft form through the divisions in threads or piecemeal. *The threads of this layer either loosely surround the bones, or are inserted more deeply into their very substance; wherever, that is to say, the bones are to be moved by the muscles; or they form close bundles, and determine themselves to the other parts of the bones: and when motions are to be continued mechanically through connexion of parts, then these bundles are termed ligaments;* of such there are multitudes that run from one bone or ginglymoid articulation to another, as in the ossa innominata, the pelvis, the false ribs,* &c. It is the disposition of the parts alone that causes the substance to appear either softish, hardish, or tender, as the case may be, and to be determined in one way, or in another, and to conduce particularly to the continued connexion of the motions. *The continuity of this layer is also evident from the anatomical examination of the skeleton. Thus we see that it is continued in this case over all the bones both of the body and the head, and also under the form of a white aponeurosis distinguishes one muscle from another; and at length proceeds to the vertebral column, and constitutes the tunica vaginalis of the spine, which*

* The author says *vertebræ*.

goes by the convex and also by the concave part of the spine all the way to the junction and articulation between the cervical vertebræ and the occiput, and there expands around the cranium and the bones of the head, being termed the pericranium, where it covers the cranium. A simple anatomical view is sufficient to shew this continuation of the membrane. It is therefore plain that its extension is exceedingly great. It takes origin behind the veil of the upper layer of the periostea, for it arises from the motive fibres. In fact, every tendon and aponeurosis was muscular once, but when the blood deserts it, it contracts, and degenerates into a tendon. *This lamina serves as a cushion for the blood-vessels and nerves, whereby they pass or flow in the freest manner into the osseous substance and the medullary matter thereof; but were it at all hard, neither the vessels nor the nerves, agreeably to the whole design of nature as constitutive or formative, would pass into the bones: for which reason, as soon as ever it is thickened and beset with humor, not only is the power of the muscle over the bones lost, but also the faculty of articulation: indeed, caries and exostoses are produced whenever this lamina is either abraded, or suffers any other lesion.* This is taught us sufficiently by medical and empirical experience: the same position is supported by reasoning on the connexion of parts; and by what takes place when the path whereby the bones are nourished is subverted and obliterated.

5. *These two lamellæ of the periosteum are perfectly distinct from each other, and indeed the innermost layer is separate or distinct from the surface of the bone, for it does not adhere to it except where tendinous or ligamentous threads, and blood-vessels and nervous fibres, are inserted into the bones. And as the laminæ are distinct, and the texture of the periosteum is lax or loose, it follows that some species of liquid permeates and circulates through it; that liquid, namely, which is expressed from the muscles, especially when in motion, and that is collected between the motive fibres, and in the cellular tissue of the muscles, which liquid is oily as well as watery, and gently touches and anoints the bones.* The motion of these layers arises from that of the surrounding muscles, which as they are variously called into play, exert a varying action upon these layers, indeed a reciprocal action between the internal and external periosteum,

causing a traction and retraction—a drawing to and fro,—whereby the liquid is driven from part to part. The extension of these membranes shews what kind of circulation there is. *According to the common rule, all the liquid tends from the circumferences to the centres, or from the greatest motion to the least; hence from the several bones, which in this case are the circumferences, to the vertebral column, which is the central axis of the bones.* The reader will find this rule confirmed throughout our pages, as also the statement that the vertebral column is the axis and central region of all the bones of the body, *i. e.*, of the whole skeleton. *And so when it reaches the vertebral column, it rises both outside and inside, and by a kind of pumping or syringe-like action is drawn up all the way to the cranium, which contains the very highest or most absolutely central part of the body.* Respecting the manner of this pumping, see our Part on the Vertebral Column, and also that on the Cranium; for thither the liquor mounts, and there its gyre terminates. *In this place, that is to say, in the cranium, this oleaginous liquor seems to be sent outwards, in the shape of effluvial halitus; it also serves for nutrition both to the hair of the head and to the bones.* On these subjects the reader is requested to wait for the Part on the Brain. As to the quantity of similar fatty and unctuous liquor that is found in the vertebral theca, and also around the pericranium, consult mere experience on the subject. *But with regard to the particular juice that exudes from the membranous filaments, it seems to be expressed at one and the same time in the very vertex or central place, or from the external periosteum, and from the internal or dura mater.* And by virtue of the same causes. This juice is the same as that which is commonly excreted from the external parts, after the residual serum has left it; and it is fluid and comparatively fatty.

6. *But the fluid that circulates through the muscular membranes and their cellular tissues, and in the end through the periostea, is discharged in divers places; that is to say, both into the adipose cells, producing fat, (for the adipose tunic lies next to the panniculus carnosus,) and in all the interstices [of the body]; and possibly it may likewise be discharged through the more open pores of the skin, and in this case pass off in sweat; and again, towards the peritonæum and the pleura, whose chan-*

nels of discharge have been already pointed out; for when the pores of the skin are shut, and perspiration prevented, it then either circulates through the periosteum, or is driven towards the peritonæum and pleura; and* vice versâ. It is well worth our while to investigate this circulation, for the health of the body very considerably depends upon it. Whenever the pores of the skin are obstructed, and the body is excited by no movements, all this excrementitious humor is thrown upon the periostea, and there made to circulate, and hence the periostea are infested by this worthless kind of serum, as may be concluded from the stuffing experienced during catarrh; or else it is sent towards the peritonæum and the pleura, the more fatty portion remaining in the adipose membrane. Thus the worthless part is always rejected, or cast out of the doors of the skin, or else sent into banishment by some other outlet; the more genuine part however is absorbed by the veins and lymphatics; the residue, which serves as an unguent, circulates through the cellular tissue and even through the periostea. This is proved by anatomical enquiries. For there are channels leading from the panniculus carnosus towards the pleura and the peritonæum; and also towards the skin and its pores; as we have proved above. Hence the practical art and science of animal nature consist in this—that the viler matter be rejected, and that the better be retained. The mode in which this is ensured will be explained in our Pathology, where the proofs on the subject can be brought forward. This equilibrium depends on the natural state of the body; and the loss of it, on the change of that state. The number of causes that go to that state, will be seen in our Rational Pathology.

* See the chapters on those membranes in *The Animal Kingdom*.—*Tr.*

FINIS.

Walton and Mitchell, Printers, Wardour-street, Oxford-street.